U.S. Department of Justice
Office of Justice Programs
National Institute of Justice

I0475761

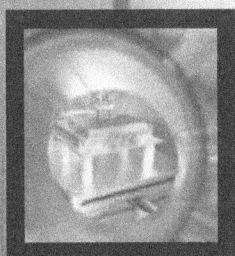

Improved Analysis of

DNA Short Tandem Repeats

With Time-of-Flight
Mass Spectrometry

science and technology research report

U.S. Department of Justice
Office of Justice Programs
810 Seventh Street N.W.
Washington, DC 20531

John Ashcroft
Attorney General

Office of Justice Programs	National Institute of Justice
World Wide Web Site	World Wide Web Site
http://www.ojp.usdoj.gov	*http://www.ojp.usdoj.gov/nij*

IMPROVED ANALYSIS OF DNA SHORT TANDEM REPEATS WITH TIME-OF-FLIGHT MASS SPECTROMETRY

John M. Butler and Christopher H. Becker

Science and Technology Research Report

October 2001
NCJ 188292

Sarah V. Hart
Director, National Institute of Justice

Lois Tully
Project Monitor

John M. Butler, Ph.D., is currently a research chemist at the National Institute of Standards and Technology and principle investigator on an NIJ-funded project to further develop multiplex PCR and time-of-flight mass spectrometry for future forensic DNA typing assays. He was the first to demonstrate that short tandem repeat typing could be performed with capillary electrophoresis.

Christopher H. Becker, Ph.D., is currently senior director of proteomics technology at Thermo Finnigan in San Jose, California. During the span of this project, he was president and chief operations officer of GeneTrace Systems, Inc.

This project was supported under grant number 97–LB–VX–0003 from the National Institute of Justice, Office of Justice Programs, U.S. Department of Justice. Points of view in this document are those of the authors and do not necessarily represent the official position or policies of the U.S. Department of Justice.

This document is not intended to create, does not create, and may not be relied upon to create any rights, substantive or procedural, enforceable at law by any party in any matter, civil or criminal.

For further information, contact John M. Butler, National Institute of Standards and Technology, 100 Bureau Drive, Gaithersburg, MD 20899; phone 301–975–4049; e-mail john.butler@nist.gov.

The National Institute of Justice is a component of the Office of Justice Programs, which also includes the Bureau of Justice Assistance, the Bureau of Justice Statistics, the Office of Juvenile Justice and Delinquency Prevention, and the Office for Victims of Crime.

ACKNOWLEDGMENTS

The project described in this report could not have happened without the hard work and support of a number of people at GeneTrace Systems, Inc. First and foremost, Jia Li did some of the early primer design and STR work to demonstrate that STRs could be effectively analyzed by mass spectrometry. Jia taught us a lot about PCR and was always encouraging of our work. Likewise, Tom Shaler was important in the early phases of this research with his expert advice in mass spectrometry and data processing. The first GeneTrace STR mass spectra were carefully collected by Tom, and thus he and Jia deserve credit for helping obtain the funding for this study. Dan Pollart synthesized numerous cleavable primers for this project, especially in the first year of our work. David Joo and Wendy Lam also prepared PCR and SNP primers for the later part of this work. A number of people assisted in robotic sample preparation and sample cleanup, including Mike Abbott, Jon Marlowe, David Wexler, and Rebecca Turincio. Joanna Hunter, Vera Delgado, and Can Nhan ran many of the STR samples on the automated mass spectrometers. Their hard work made it possible to focus on experimental design and data analysis rather than routine sample handling.

It was a great blessing to have talented and supportive coworkers throughout the course of this project. Kathy Stephens, Jia Li, Tom Shaler, Yuping Tan, Christine Loehrlein, Joanna Hunter, Hua Lin, Gordy Haupt, and Nathan Hunt provided useful discussions on a number of issues and helped develop assay parameters and tackle automation issues, among other things. Nathan Hunt was especially important to the success of this project because he developed the STR genotyping algorithm and CallSSR software as well as the multiplex SNP primer design software. Kevin Coopman developed the SNP genotyping algorithm and calling software and was always eager to analyze our multiplex SNP samples. Joe Monforte and Roger Walker served as our supervisors for the first year and second year of this project, respectively, which allowed us the opportunity to devote sufficient time to doing the work described in this project. Last but not least, Debbie Krantz served as an able administrator of these two NIJ grants and took care of the financial aspects.

We also were supported with samples and sequence information from a number of scientific collaborators. Steve Lee and John Tonkyn from the California Department of Justice DNA Laboratory provided genomic DNA samples and STR allelic ladders. Debang Liu from Northwestern University provided the D3S1358 DNA sequence used for improved primer design purposes. Peter Oefner and Peter Underhill from the Department of Genetics at Stanford University provided male population samples and Y-chromosome SNP sequences. The encouragement and support of Lisa Forman and Richard Rau from the Office of Justice Programs at the National Institute of Justice propelled this work from an idea to a working product. In addition, Dennis Reeder from the National Institute of Standards and Technology was always a constant source of encouragement at scientific meetings.

CONTENTS

List of exhibits included with this report:

EXECUTIVE SUMMARY

Introduction

The advent of DNA typing and its use for human identity testing has revolutionized law enforcement investigations in recent years by allowing forensic laboratories to match suspects with minuscule amounts of biological evidence from a crime scene. Equally important is the use of DNA to exclude suspects who were not involved in a crime or to identify human remains in an accident.

The past decade has seen numerous advances in the DNA testing procedures, most notably among them the development of PCR (polymerase chain reaction)-based DNA typing methods. Technologies for measuring DNA variations, both length and sequence polymorphisms, have also advanced rapidly in the past decade. The time needed to determine a sample's DNA profile has dropped from 6–8 weeks to 1–2 days, and with more recent advancements, the time needed to process samples may decrease to as little as a few hours, maybe even a few minutes.

Simultaneous with the evolution of DNA markers and technologies embraced by the forensic community has been the acceptance and use of DNA typing information. The courtroom battles over statistical issues that were common in the late 1980s and early 1990s have subsided as DNA evidence has become more widely accepted.

In the past 5 years, DNA databases have emerged as powerful tools for criminal investigations, much like the fingerprint databases that have been used routinely for decades.

The United Kingdom launched a nationwide DNA database in 1995 that now contains more than 1 million DNA profiles of convicted felons—profiles that have been used to aid more than 75,000 criminal investigations. National DNA databases are springing up in countries all over the world as their value to law enforcement is being recognized.

In the United States, the FBI has developed the Combined DNA Index System (CODIS) with the anticipation that several million DNA profiles will be entered into this database in the next decade. All 50 States now have laws requiring DNA typing of convicted offenders, typically for violent crimes such as rape or homicide.

While the law enforcement community is gearing up to gather millions of DNA samples from convicted felons, the DNA typing technology needs improvement. Large backlogs of samples exist today due to the high cost of performing the DNA testing and limited capabilities in forensic laboratories. As of the summer of 1999, several States, including California, Virginia, and Florida, had backlogs of more than 50,000 samples. A need exists for more rapid and cost-effective methods for high-throughput DNA analysis to process samples currently being gathered for large criminal DNA databases around the world.

At the start of this project in June 1997, commercially available slab gel or capillary electrophoresis instruments could handle only a few dozen samples per day. While larger numbers of samples can be processed by increasing the number of laboratory personnel and instruments, the development of high-throughput DNA processing technologies promises to be more cost effective in the long run, especially for the generation of large DNA databases. GeneTrace Systems, Inc., a small biotechnology company located in Alameda, California, has developed high-throughput DNA analysis capabilities using time-of-flight mass spectrometry coupled with parallel sample preparation on a robotic workstation. The GeneTrace technology allows several thousand samples to be processed daily. DNA samples can be analyzed in seconds, rather than minutes or hours, and with improved accuracy compared with conventional electrophoresis methods.

Overall, the mass spectrometry method described in this study is two orders of magnitude faster in sample processing time than conventional techniques.

Purpose of the Report

This NIJ project was initiated to adapt the GeneTrace technology to human identity DNA markers commonly used by forensic DNA laboratories, specifically short tandem repeat (STR) markers. An extension of the original grant was submitted in December 1997 to fund the development of single nucleotide polymorphism (SNP) markers from mitochondrial DNA and the Y chromosome.

Based on the results obtained in this study, the authors believe mass spectrometry can be a useful and effective means for high-throughput DNA analysis, and that it has the capabilities to meet the needs of the forensic DNA community for offender DNA databases.

However, due to limited resources and a perceived difficulty to enter the forensic DNA market, GeneTrace made a business decision to not pursue this market. While the STR milestones on the original grant were met, only the initial milestones were achieved on the SNP portion of the NIJ grant because of the premature termination on the part of GeneTrace.

GeneTrace Systems, Inc., developed an integrated high-throughput DNA analysis system involving the use of proprietary chemistry, robotic sample manipulation, and time-of-flight mass spectrometry. The purpose of this NIJ project was to apply the GeneTrace technology to improve the analysis of STR markers commonly used in forensic DNA laboratories.

Mass spectrometry is a versatile analytical technique that involves the detection of ions and the measurement of their mass-to-charge ratio. Because these ions are separated in a vacuum environment, analysis times can be extremely rapid, often within microseconds. Many advances have been made in the past decade for the analysis of biomolecules such as DNA, proteins, and carbohydrates since the introduction of a new ionization technique known as matrix-assisted laser desorption-ionization (MALDI) and the discovery of new matrixes that effectively ionize DNA without extensive fragmentation. When coupled with time-of-flight mass spectrometry, this method for measuring biomolecules is commonly referred to as MALDI-TOF-MS. A schematic of MALDI-TOF-MS is presented in exhibit 1.

Short Tandem Repeats

Short tandem repeat (STR) DNA markers, also referred to as microsatellites or simple sequence repeats (SSRs), consist of tandemly repeated DNA sequences with a core repeat of 2–6 base pairs (bp). STR markers are readily amplified during PCR by using primers that bind in conserved regions of the genome flanking the repeat region. Forensic laboratories prefer tetranucleotide loci (i.e., 4 bp in the repeat) due to the lower amount of "stutter" produced during PCR. (Stutter products are additional peaks that can complicate the interpretation of DNA mixtures by appearing in front of regular allele peaks.) The number of repeats can vary from 3 or 4 repeats to more than 50 repeats with extremely polymorphic markers. The number of repeats, and hence the size of the PCR product, may vary among samples in a population making STR markers useful in identity testing or genetic mapping studies.

Shortly after this project was initiated, the FBI designated 13 core STR loci for the nationwide CODIS database. These STR loci are TH01, TPOX, CSF1PO, VWA, FGA, D3S1358, D5S818, D7S820, D13S317, D16S539, D8S1179, D18S51, and D21S11. The sex-typing marker, amelogenin, is also included in STR multiplexes that cover the 13 core STR loci. Each sample must have these 14 markers tested to be entered into CODIS.

To illustrate the kinds of numbers involved to analyze the current national sample backlog of ~500,000 samples, more than 7 million genotypes must be generated. Using currently available technologies, an estimated $25 million (~$50/sample) and more than 5 years for well-trained and well-funded

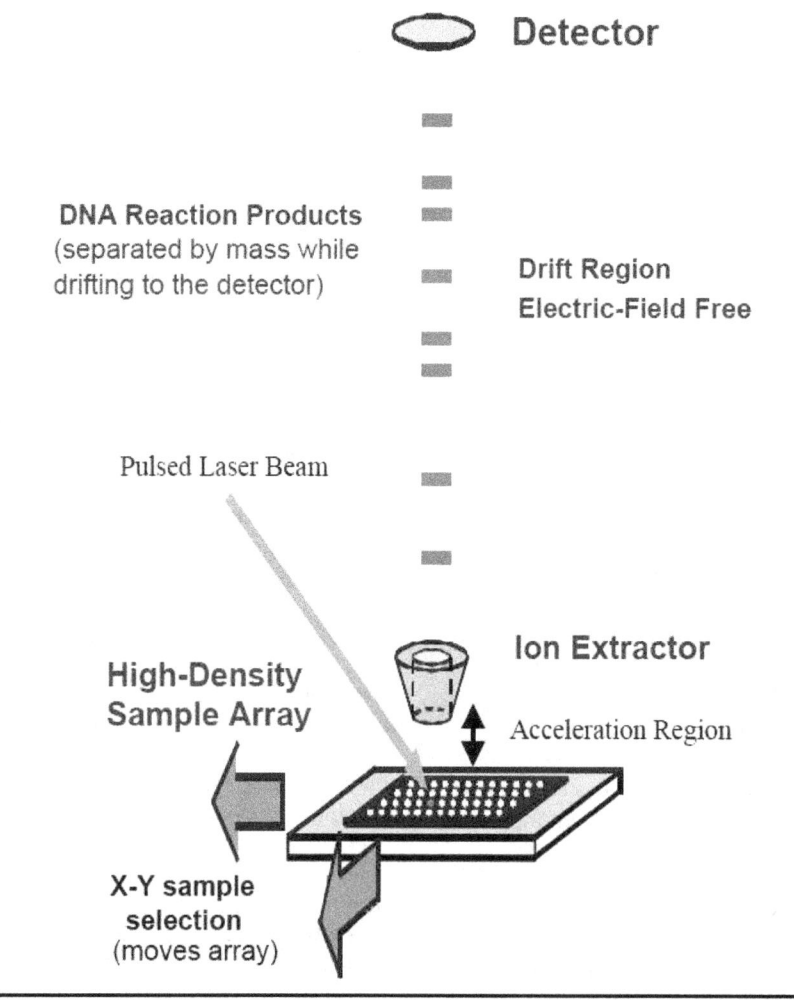

Exhibit 1. **Schematic of GeneTrace automated time-of-flight mass spectrometer.** DNA molecules are liberated from a solid-phase matrix environment with a laser pulse. The DNA reaction products are separated by size (mass) in a matter of microseconds, as opposed to hours using conventional methods. For each run, hundreds of samples are prepared in parallel using a robotic workstation and spotted on a sample plate that is introduced to the vacuum environment of the mass spectrometer. The sample plate moves under the fixed laser beam to allow sequential sample analysis.

Detector

DNA Reaction Products
(separated by mass while drifting to the detector)

**Drift Region
Electric-Field Free**

Pulsed Laser Beam

Ion Extractor

**High-Density
Sample Array**

Acceleration Region

**X-Y sample
selection**
(moves array)

laboratories would be required to determine those 7 million genotypes. With the high cost and effort required, most of these backlogged samples are being stored in anticipation of future analysis and inclusion in CODIS, pending the development of new, faster technology or the implementation of more instruments using the current electrophoresis technologies.

Exhibit 2. **PCR product sizes with newly designed primers for commonly used STR loci compared with commercially available primers used in multiplex sets for fluorescence-based assays.** The red numbers indicate PCR product size ranges that exceed the recommended 140 bp mass spectrometry detection range.

STR Locus	Known Alleles	GeneTrace Sizes (newly designed primers in this study)	Commerically Available Sizes*	
			Applied Biosystems	Promega
Amelogenin	X, Y	106, 112 bp	106, 112 bp	212, 218 bp
CD4	4–15	81–136 bp	N.A.	
CSF1PO	6–15	87–123 bp	280–316 bp	291–327 bp
F13A1	3–17	112–168 bp	N.A.	279–335 bp
F13B	6–12	110–134 bp	N.A.	169–193 bp
FES/FPS	7–15	76–108 bp	N.A.	222–254 bp
FGA	15–30	118–180 bp	206–266 bp	N.A.
D3S1358	9–20	76–120 bp	101–145 bp	N.A.
D5S818	7–15	89–121 bp	134–166 bp	119–151 bp
D7S820	6–14	66–98 bp	257–289 bp	215–247 bp
D8S1179	8–18	92–130 bp	127–167 bp	N.A.
D13S317	7–15	98–130 bp	201–233 bp	165–197 bp
D16S539	5,8–15	81–121 bp	233–273 bp	264–304 bp
D18S51	9–27	120–192 bp	272–344 bp	N.A.
D21S11	24–38	150–190 bp	186–242 bp	N.A.
DYS19	8–16	76–108 bp	N.A.	
DYS391	9–12	99–111 bp	N.A.	
HPRTB	6–17	84–128 bp	N.A.	259–303 bp
LPL	7–14	105–133 bp	N.A.	105–133 bp
TH01	3–13.3	55–98 bp	160–203 bp	171–214 bp
TPOX	6–14	69–101 bp	217–249 bp	224–256 bp
VWA	11–22	126–170 bp	156–200 bp	127–171 bp
Other STRs				
GATA132B04	10–14	99–115 bp	N.A.	
D22S445	10–16	110–130 bp	N.A.	
D16S2622	4–8	71–87 bp	N.A.	

* Sizes are listed without adenylation (add 1 base for +A form).
N.A. = Not Available.

Time-of-flight mass spectrometry has the potential to bring DNA sample processing to a new level in terms of high-throughput analysis. However, there are several challenges to using MALDI-TOF-MS for the analysis of PCR products, such as STR markers. Mass spectrometry resolution and sensitivity are diminished when either the DNA size or the salt content of the sample is too large. By redesigning the PCR primers to bind close to the repeat region, the STR allele sizes are reduced to benefit the resolution and sensitivity of the PCR products. Therefore, much of this project involved designing and testing new PCR primers that produced smaller amplicon sizes for STR markers of forensic interest. This research focused on STR loci that have been developed by commercial manufacturers and studied extensively by forensic scientists. These include all of the GenePrint™ tetranucleotide STR systems from Promega Corporation (Madison, WI) as well as the 13 CODIS STR loci that are covered by the Profiler Plus™ and COfiler™ kits from Applied Biosystems (ABI) (Foster City, CA) (exhibit 2). Where possible, primers were designed to produce amplicons less than 100 bp in size, although it has been possible to resolve neighboring STR alleles as large as 140 bp. For example, TPOX alleles 6–14 ranged from 69–101 bp in size with GeneTrace-designed primers; while with Promega's GenePrint™ primers, the same TPOX alleles ranged in size from 224–256 bp. Unfortunately, due to the long and complex repeat structures of several STR markers, this study was unable to obtain the necessary single-base resolution with the following STR loci: D21S11, D18S51, and FGA (see Results and Discussion of STR Analysis by Mass Spectrometry).

To verify the STR results obtained from the mass spectrometry method, the authors collaborated with the California Department of Justice (CDOJ) DNA Laboratory in Berkeley to generate a large data set. CDOJ provided 88 samples that had been previously genotyped using validated fluorescent multiplex STR kits from ABI. GeneTrace generated STR results using their primer sets for 9 STR loci (TH01, TPOX, CSF1PO, D3S1358, D16S539, D8S1179, FGA, DYS391, and D7S820) along with the sex-typing marker amelogenin. These experiments compared more than 700 genotypes (88 samples X 8 loci; data from D8S1179 and DYS391 were not available from CDOJ). Although results were not obtained for all possible samples using mass spectrometry, researchers observed almost 100% correlation with the genotypes obtained between the validated fluorescent STR method and GeneTrace's newly developed mass spectrometry technique, demonstrating that the GeneTrace method was reliable (see Results and Discussion of STR Analysis by Mass Spectrometry).

Multiplex STR analysis

To reduce analysis cost and sample consumption and to meet the demands of higher sample throughputs, PCR amplification and detection of multiple markers (multiplex STR analysis) has become a standard technique in most forensic DNA laboratories. STR multiplexing is most commonly performed using spectrally distinguishable fluorescent tags and/or nonoverlapping PCR product sizes. An example of an STR multiplex produced from a commercially available kit is shown in exhibit 3. Multiplex STR amplification in one or two PCR reactions with fluorescently labeled primers and measurement with gel or capillary electrophoresis separation and laser-induced fluorescence detection is becoming a standard method among forensic laboratories for analysis of the 13 CODIS STR loci. The STR alleles from these multiplexed PCR products typically range in size from 100–350 bp with commercially available kits.

Due to the limited DNA size constraints of mass spectrometry, GeneTrace adopted a different approach to multiplex analysis of multiple STR loci. Primers

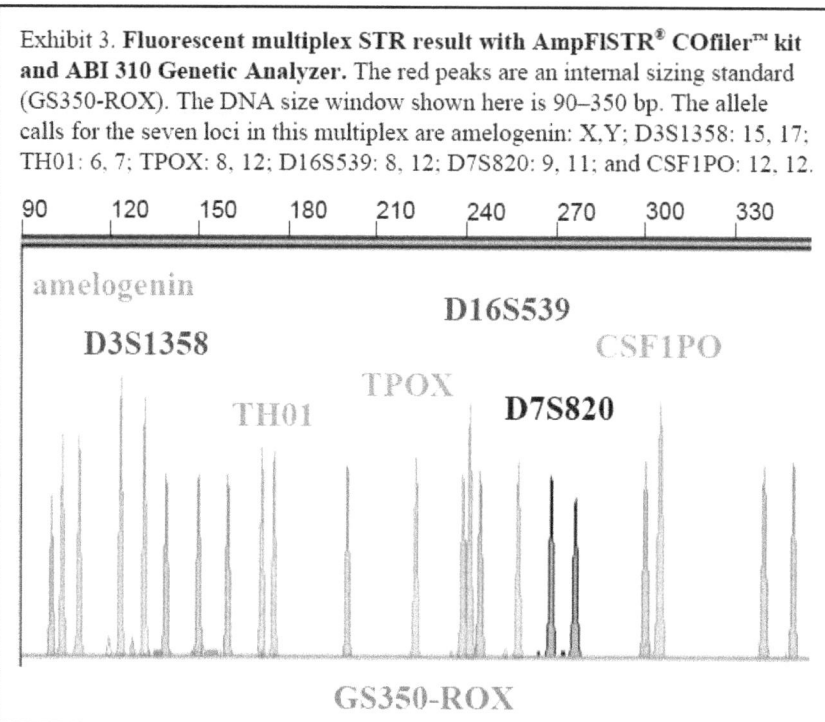

Exhibit 3. **Fluorescent multiplex STR result with AmpFlSTR® COfiler™ kit and ABI 310 Genetic Analyzer.** The red peaks are an internal sizing standard (GS350-ROX). The DNA size window shown here is 90–350 bp. The allele calls for the seven loci in this multiplex are amelogenin: X,Y; D3S1358: 15, 17; TH01: 6, 7; TPOX: 8, 12; D16S539: 8, 12; D7S820: 9, 11; and CSF1PO: 12, 12.

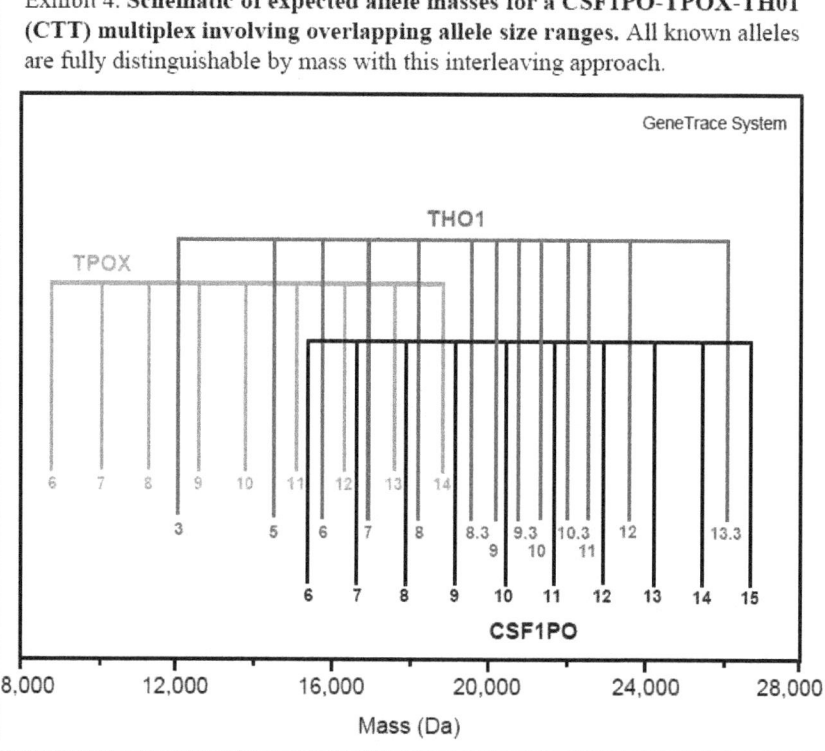

Exhibit 4. **Schematic of expected allele masses for a CSF1PO-TPOX-TH01 (CTT) multiplex involving overlapping allele size ranges.** All known alleles are fully distinguishable by mass with this interleaving approach.

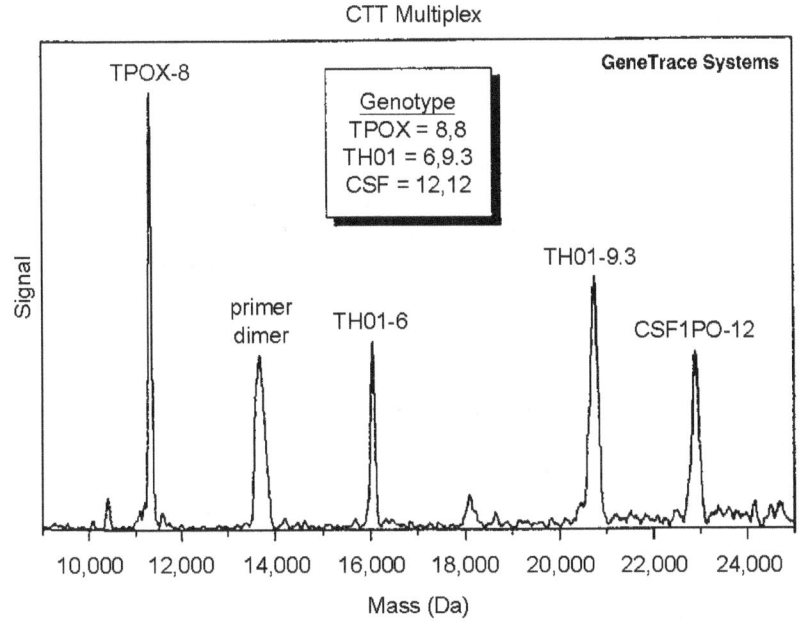

Exhibit 5. **Mass spectrum of an STR triplex involving TPOX, TH01, and CSF1PO.** This sample is a mass spectrometry result using the interleaving allele approach schematically illustrated in exhibit 4. Multiplex PCR and multiplex primer extension with ddC termination were used to obtain this result.

are designed such that the PCR product size ranges overlap between multiple loci but have alleles that interleave and are resolvable in the mass spectrometer (exhibit 4). As described above, PCR primers are closer to the STR repeat regions than those commonly used with electrophoresis systems. The high accuracy, precision, and resolution of this mass spectrometry approach permits multiplexing STR loci for a limited number of markers. During the study, GeneTrace also developed a TH01-TPOX-CSF1PO STR triplex (exhibit 5).

Single Nucleotide Polymorphisms

Single nucleotide polymorphisms (SNPs) represent another form of DNA variation that is useful for human identity testing. SNPs are the most frequent form of DNA sequence variation in the human genome and are becoming increasingly popular genetic markers for genome mapping studies and medical diagnostics. SNPs are typically biallelic with two possible nucleotides (nt) or alleles at a particular site in the genome. Because SNPs are less polymorphic (i.e., have fewer alleles) than the currently used STR markers, more SNP markers are required to obtain the same level of discrimination between samples. Current estimates are that 30–50 unlinked SNPs will be required to obtain the matching probabilities of 1 in ~100 billion as seen with the 13 CODIS STRs.

The perceived value of SNPs for DNA typing in a forensic setting include the following:

◆ More rapid analysis.

◆ Cheaper costs.

◆ Simpler interpretation of results because there are no stutter products.

◆ Improved ability to handle degraded DNA because of the possibility of smaller PCR product sizes.

While it is doubtful that autosomal SNPs will replace the current battery of STRs used in forensic laboratories in the near future, abundant mitochondrial and Y-chromosome SNP markers exist and have already proven useful as screening tools. These maternal (mitochondrial) and paternal (Y chromosome) lineage markers are effective in identifying missing persons and war casualties and helping answer historical questions such as whether or not Thomas Jefferson fathered a slave child.

The forensic DNA community already has experience with applying SNP markers as a screening process, which can prove very helpful for excluding suspects from crime scenes. Many crime laboratories still use reverse dot blot technology for analyzing the SNPs from HLA-DQA1 and PolyMarker loci with kits from ABI. In addition, mitochondrial DNA (mtDNA) sequencing is currently performed in some forensic laboratories.

In the work performed on multiplex SNP markers at GeneTrace, the authors examined 10 polymorphic sites within the mtDNA control region and 20 Y-chromosome SNPs provided by Dr. Peter Oefner and Dr. Peter Underhill from Stanford University. A multiplex SNP assay was developed for 10 mtDNA SNP sites (exhibit 6). Only limited work was performed on the Y-chromosome SNPs due to the premature termination of the work. However, results demonstrated a male-specific 17-plex PCR of 17 different Y SNP markers (exhibit 7).

Conclusions and Implications

Time-of-flight mass spectrometry offers a rapid, cost-effective alternative for genotyping large numbers of samples. Each DNA sample can be accurately measured in a few seconds. Due to the increased accuracy of mass spectrometry, STR alleles can be reliably typed without comparison with allelic ladders. Mass spectrometry holds significant promise as a technology for high-throughput DNA processing that will be valuable for large-scale DNA database work.

In summary, the positive features of mass spectrometry for STR analysis include:

◆ Rapid results—STR typing at a rate of seconds per sample.

◆ Accuracy—no allelic ladders.

◆ Direct DNA measurement—no fluorescent or radioactive labels.

◆ Automated sample preparation and data collection.

◆ High-throughput capabilities of thousands of samples daily per system.

◆ Flexibility—single nucleotide polymorphism (SNP) assays can be run on the same instrument platform.

This project demonstrates that both STR and SNP analysis are reliably performed with GeneTrace's mass spectrometry technology. Tests were done on a large number of human DNA markers of forensic interest. New primer sets were developed for the 13 CODIS STR loci that may prove useful in the future for situations in which degraded DNA is present and requires smaller amplicons to obtain

successful results. The possibility of developing multiplexed SNP markers also was explored, and a mtDNA 10-plex assay and Y-chromosome, 17-plex, male-specific PCR were demonstrated. Both STR and SNP areas appear promising for future research. In another project, GeneTrace recently demonstrated a sample throughput of approximately 4,000 STR samples in a single day with a single automated mass spectrometer. Clearly, this is an improvement in the analysis of DNA short tandem repeat markers using time-of-flight mass spectrometry.

Exhibit 6. **Mass spectrum of SNP 10-plex assay for screening polymorphic sites in the mtDNA control region.** The bottom panel shows the 10 SNP primers prior to the primer extension reaction. The top panel contains the multiplexed reaction products, each labeled with the observed extension product. The results for this K562 PCR product are (in order across HV1 and HV2): H16069 (G), H16129 (C), H16189 (A), H16224 (G), H16311 (A), H00073 (C), L00146 (T), H00152 (A), L00195 (T), and H00247 (C). SNP nucleotide results have been confirmed by sequencing. The primer sequences are listed in exhibit 24. The primer concentrations were 25 pmol MT5, 15 pmol MT8', 15 pmol MT10, 10 pmol MT3', 20 pmol MT9, 25 pmol MT6, 20 pmol MT2g, 27 pmol MT1, 35 pmol MT7, and 20 pmol MT4e.

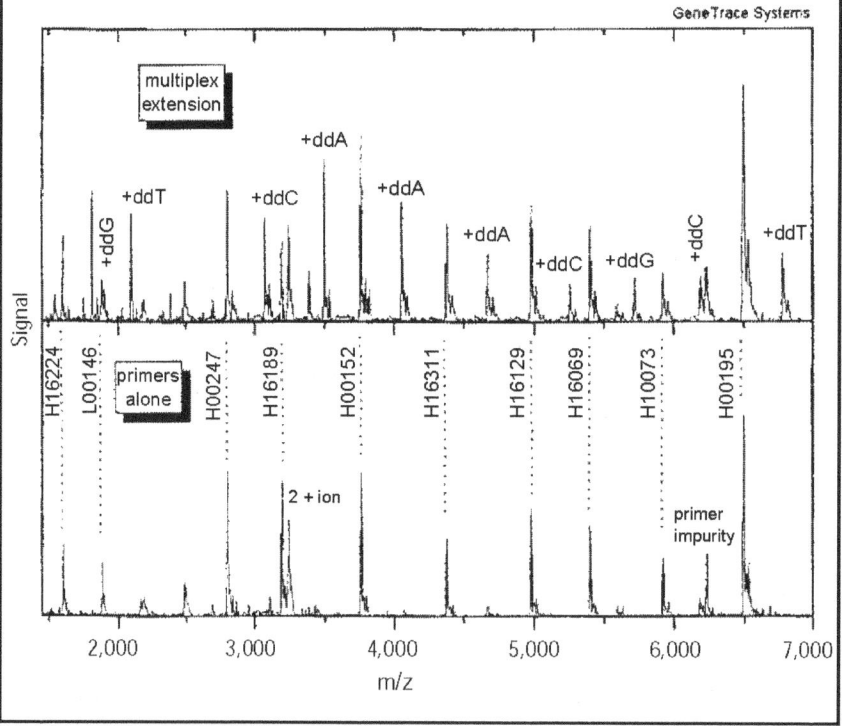

Exhibit 7. **Overlay of CE electropherograms demonstrating male-specific amplification with the 17-plex set of PCR primers.** Note: The female sample (K562) failed to yield any peaks, illustrating that the PCR reaction is Y chromosome-specific. The PCR primers used are described in exhibit 27 with each locus-specific primer set at 0.4 pmole each. For the PCR reaction conditions, see exhibit 67 and details in the text. The 17 amplicons may be seen more clearly in exhibit 68.

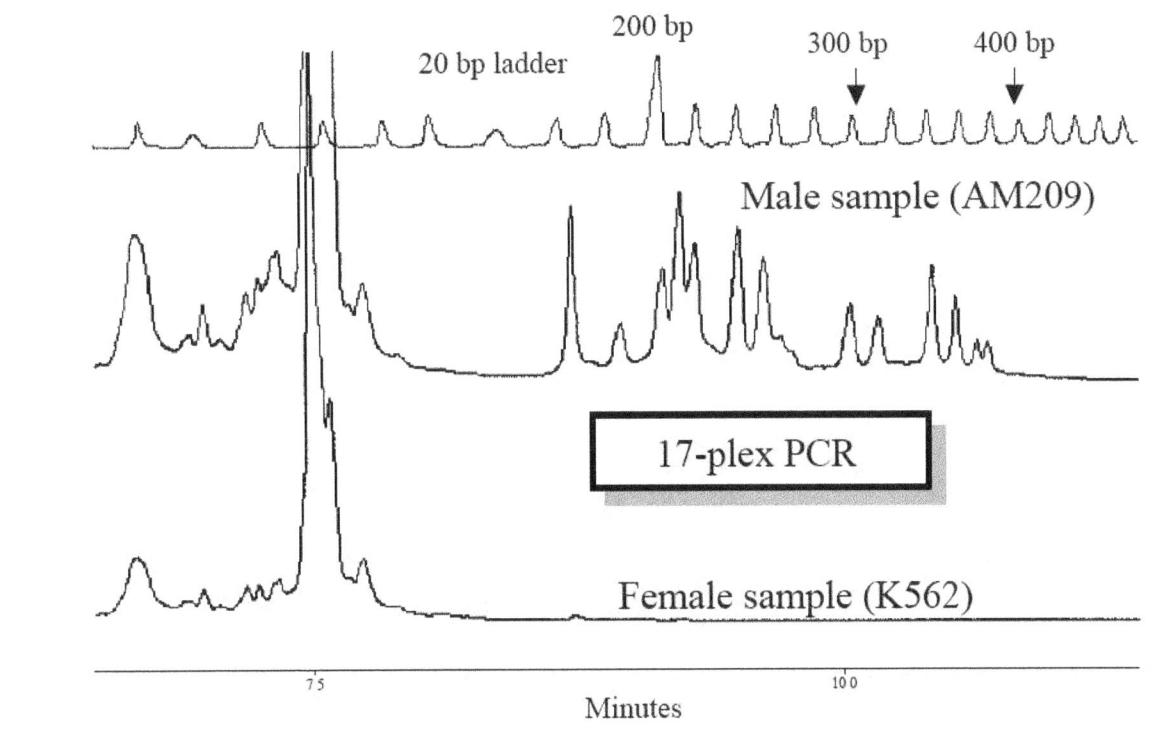

PROJECT DESCRIPTION

STR Grant

This project focused on the development of a powerful new technology for rapid and accurate analysis of DNA STR markers using time-of-flight mass spectrometry. GeneTrace Systems, Inc., collaborated with the CDOJ DNA Laboratory in Berkeley, California, primarily through Dr. Steve Lee. This collaboration provided the study with the samples used to verify the new GeneTrace technology, which was done by comparing the mass spectrometry results with genotypes obtained using established and validated methods run at CDOJ.

To accomplish the task of developing a new mass spectrometry technology for STR typing, five milestones were proposed in the original grant application, which included the following:

◆ Redesign PCR primers for a number of commonly used STR markers to produce smaller PCR products that could be tested in the mass spectrometer (exhibit 2).

◆ Demonstrate multiplexing capabilities to a level of 2 or 3 for detection with the TOF-MS method (exhibits 4–5 and 8–10).

◆ Transfer the sample preparation protocols from manual to a highly parallel and automated pipetting robot.

◆ Develop a large data set to confirm the accuracy and reliability of this method (exhibits 11–19).

◆ Automate and incorporate DNA extraction techniques onto the GeneTrace robots.

As described in the results section, all milestones were met on time except the final one regarding DNA extraction. Two other companies, Rosys and Qiagen, produced robotic systems for DNA extraction after this project began. Meanwhile, GeneTrace remained focused on developing

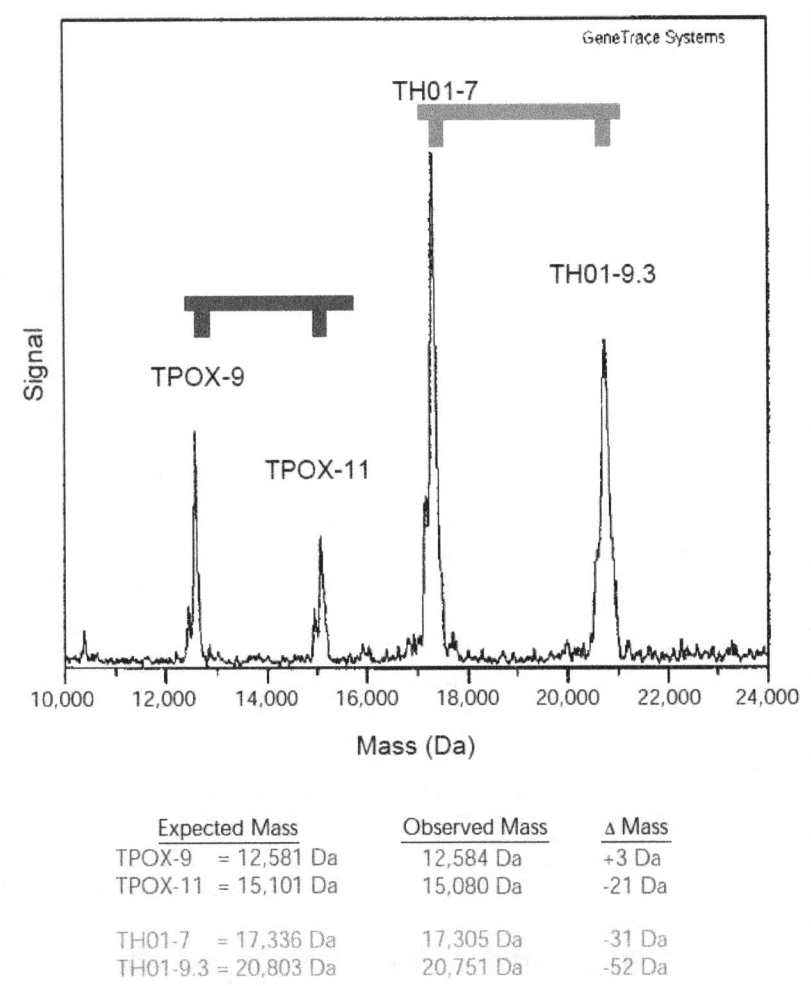

Exhibit 8. **Mass spectrum of an STR multiplex sample with nonoverlapping alleles.** The two loci, TH01 and TPOX, were coamplified using the ddC termination approach. Note: The mass accuracy is improved for peaks closer to the calibration standard of 10,998 Da.

Expected Mass	Observed Mass	Δ Mass
TPOX-9 = 12,581 Da	12,584 Da	+3 Da
TPOX-11 = 15,101 Da	15,080 Da	-21 Da
TH01-7 = 17,336 Da	17,305 Da	-31 Da
TH01-9.3 = 20,803 Da	20,751 Da	-52 Da

other steps in DNA sample processing since commercially available solutions had already been developed, thereby eliminating the need to include the DNA extraction portion in this study.

Since this project began in June 1997, a number of advances that impact the ability to perform high-throughput DNA typing have occurred in the biotech field. In early 1998, ABI released a dual 384-well PE9700 ("Viper") thermal cycler, which makes it possible to prepare 768 PCR samples simultaneously. Beckman Instruments

also came to market with a 96-tip Multimek pipetting robot. At the beginning of this project, GeneTrace used funds from this NIJ grant to purchase an MJ Research 384-well thermal cycler and a custom-built 96-tip robotic pipettor on a CyberLab x-y-z gantry. Both of these pieces of equipment were the state of the art at the time but are now obsolete at GeneTrace for routine operations and have been replaced by the newer and more reliable products from Applied Biosystems and Beckman Instruments.

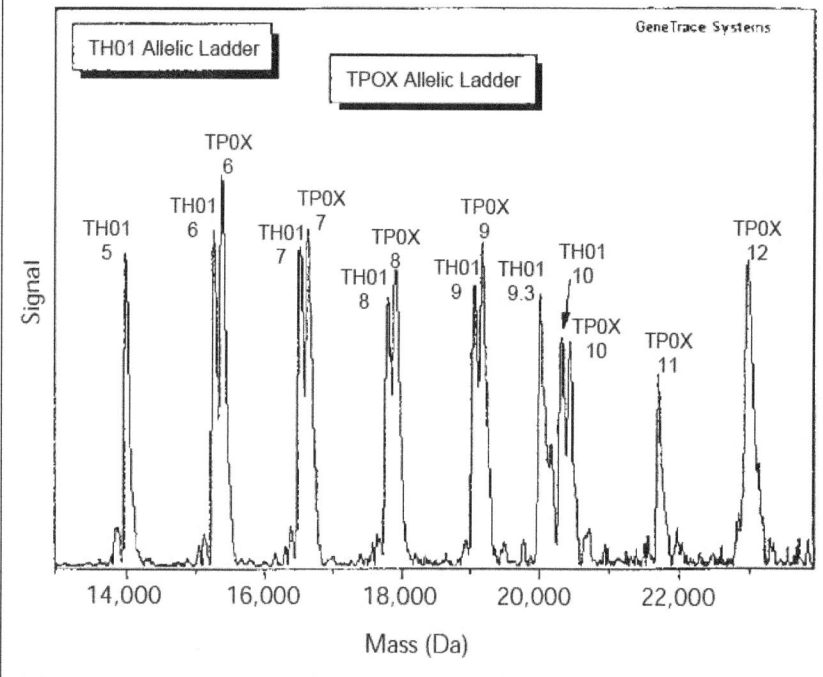

Exhibit 9. **Mass spectrum of STR multiplex mixture of TH01 and TPOX allelic ladders.** The PCR products from the two loci differ by 120 Da. The ladders were reamplified from AmpF1STR® Green I kit materials. The TH01 ladder ranges from 5 to 10 repeats with 9.3 included; the TPOX ladder ranges from 6 to 13 repeats with 13 not shown here.

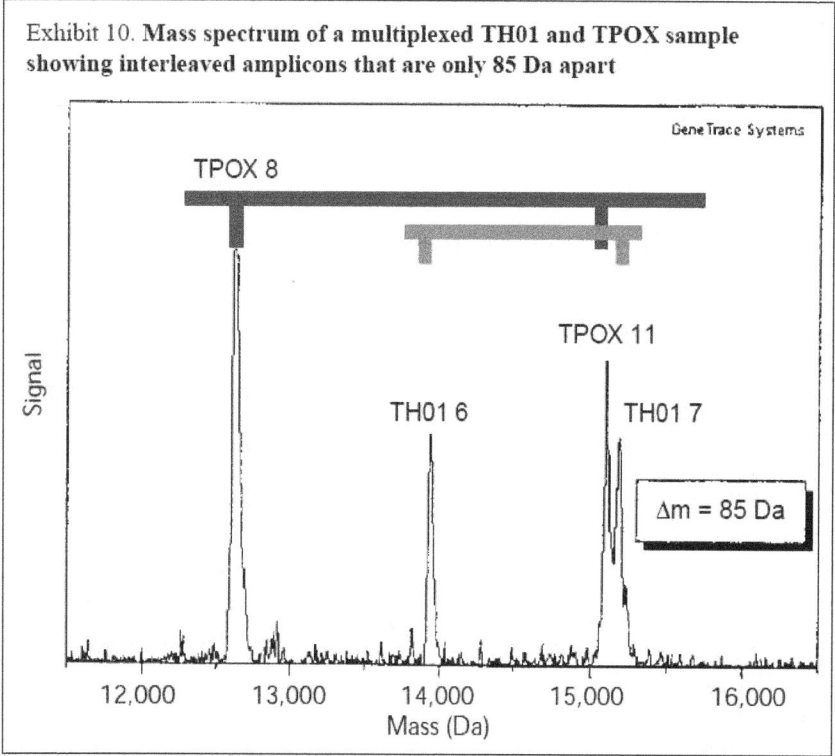

Exhibit 10. **Mass spectrum of a multiplexed TH01 and TPOX sample showing interleaved amplicons that are only 85 Da apart**

SNP Grant

The grant extension, which began in August 1998, focused on the development of multiplexed SNP markers from mtDNA and the Y chromosome. Although the grant extension was terminated prematurely by GeneTrace management in April 1999, portions of the first four milestones were accomplished. The five milestones described in the original grant extension included the following:

◆ Produce and test a set of 10 or more SNP probes for mtDNA control region "hot spots" (exhibit 5).

◆ Develop software for multiplex SNP analysis and data interpretation.

◆ Examine individual Y-chromosome SNP markers.

◆ Develop multiplex PCR and multiplex SNP probes for Y-chromosome SNP loci (exhibit 7).

◆ Determine the discriminatory power for a set of Y-chromosome markers by running ~300 samples across 50 Y-chromosome SNP markers.

The goal of the grant extension project was to develop highly multiplexed SNP assays that worked in a robust manner with mass spectrometry and could be genotyped in an automated fashion. GeneTrace planned to select markers with a high degree of discrimination to aid in rapid screening of mitochondrial DNA and Y-chromosome polymorphisms with the capability to handle analysis of large databases of offender DNA. At the time this proposal was written (December 1997), GeneTrace still intended to provide reagents and instruments to large DNA service laboratories or to provide a DNA typing service to the forensic DNA community. The grant extension was prematurely terminated due to a change in business focus and a need to consolidate the research efforts at GeneTrace.

Exhibit 11. **STR data collection times for CDOJ samples.** Comparing ABI 310 run times with multiple mass spectrometry runs indicates that data collection speed increased almost 10-fold during 1998 due to a number of improvements in the GeneTrace process. Data collection speed is a combination of laser rate, total number of shots taken, sample cleanliness, and other factors. Overall, the mass spectrometry data collection speed is approximately two orders of magnitude faster than the ABI 310. Exhibits 12–19, 39, and 40 list the observed masses for these samples measured by GeneTrace's mass spectrometry method.

ABI 310 Runs

January 5, 1999–January 8, 1999:

- Multiplex STR detection and analysis (COfiler[TM] kit amplifies 6 STRs + amelogenin simultaneously: TH01, TPOX, CSF1PO, D3S1358, D16S539, D7S820)

- ~54 hours of continuous operation (107 injections that included 88 samples, 9 allelic ladders, and reinjections of 10 samples at ~30 min per injection)

- ~2209 sec/sample or 316 sec/genotype (~5 min/genotype)

Mass Spectrometry Runs

Date	STR Loci Tested	Number of samples	Time Required for Data Collection	Average Time per Sample
Feb 11, 1998	TH01	88 + 8 controls	76 min for 96 samples	~50 sec/sample
May 6, 1998	TH01	6 samples	2.5 min for 6 samples	~25 sec/sample
October 1, 1998	TPOX, CSF1PO, D3S1358, D7S820	72 + 8 controls for each locus	60 min for 320 samples	11.3 sec/sample
Jan 12, 1999	TH01, TPOX, CSF1PO, D7S820	88 + 8 controls for each locus	30 min for 384 samples	4.7 sec/sample
Jan 14, 1999	TH01, D16S539, D3S1358, Amelogenin	88 + 8 controls for each locus	37 min for 384 samples	5.8 sec/sample
Feb 12, 1999	D3S1358, FGA, D8S1179, DYS391	88 + 8 controls for each locus	41 min for 384 samples	6.4 sec/sample
March 26, 1999	TH01, TPOX, CSF1PO, D16S539	88 + 8 controls for each locus	51 min for 384 samples	8.0 sec/sample

Exhibit 12. **CDOJ CSF1PO results compared with ABI 310 and mass spectrometry methods**

Position	ABI 310	Mass Spec	Allele 1 (Da)	Allele 2 (Da)	Position	ABI 310	Mass Spec	Allele 1 (Da)	Allele 2 (Da)
A1	11,12	11,12	27,823	29,060	A7	11,12	11,12	27,867	29,000
B1	10,12	10,12	26,520	29,033	B7	7,13	7,13	22,809	30,271
C1	7,10	7,10	22,917	26,684	C7	10,13	10,13	26,529	30,253
D1	10,13	10,13	26,586	30,344	D7	10,11	10,11	26,621	27,794
E1	8,12	8,12	24,077	29,075	E7	10,11	10,11	26,609	27,834
F1	8,10	8,10	24,107	26,639	F7	7,12	7,12	22,923	29,202
G1	12,13	12,13	29,144	30,373	G7	11,12	11,12	27,958	29,244
H1	11,12	11,12	27,812	29,018	H7	10,12	10,12	26,497	29,005
A2	12,12	12,12	29,033		A8	12,12	12,12	29,129	
B2	10,10	10,10	26,594		B8	8,11	8,11	24,117	27,862
C2	10,10	10,10	26,550		C8	7,12	7,12	22,913	29,071
D2	10,11	10,11	26,609	27,823	D8	10,10	10,10	26,575	
E2	10,12	10,12	26,620	29,129	E8	10,12	10,12	26,552	29,007
F2	10,10	10,10	26,584		F8	8,10	8,10	24,061	26,572
G2	11,12	11,12	27,862	29,080	G8	12,12	12,12	29,064	
H2	10,10	10,10	26,605		H8	11,12	11,12	27,796	29,026
A3	12,12	12,12	28,987		A9	8,12	8,12	24,085	29,091
B3	10,13	10,13	26,614	30,380	B9	11,12	11,12	27,819	29,022
C3	10,10	10,10	26,618		C9	11,12	11,12	27,810	29,000
D3	11,11	11,11	27,823		D9	8,13	8,13	24,093	30,348
E3	11,12	11,12	27,817	28,998	E9	11,12	11,12	27,878	29,098
F3	11,12	11,12	27,871	29,093	F9	7,12	7,12	22,836	29,104
G3	10,11	10,11	26,616	27,812	G9	9,13	9,13	25,387	30,389
H3	11,12	11,12	27,849	29,024	H9	7,11	7,11	22,866	27,836
A4	8,10	8,10	24,048	26,499	A10	9,10	9,10	25,391	26,607
B4	11,13	11,13	27,941	30,412	B10	11,12	11,12	27,844	29,044
C4	11,11	11,11	27,873		C10	10,10	10,10	26,586	
D4	10,11	10,11	26,638	27,794	D10	8,11	8,11	24,091	27,838
E4	7,10	7,10	22,835	26,553	E10	11,12	11,12	27,843	29,040
F4	10,14	10,14	26,592	31,565	F10	10,12	10,12	26,607	29,111
G4	12,14	12,14	29,113	31,625	G10	8,11	8,11	24,029	27,807
H4	9,10	9,10	25,360	26,582	H10	8,10	8,10	24,099	26,609
A5	10,10	10,10	26,590		A11	12,12	12,12	29,011	
B5	7,11	7,11	22,818	27,838	B11	7,12	7,12	22,836	29,095
C5	10,10	10,10	26,628		C11	7,11	7,11	22,826	27,834
D5	10,11	10,11	26,647	27,925	D11	10,11	10,11	26,567	27,784
E5	11,12	11,12	27,847	29,031	E11	10,12	10,12	26,577	29,078
F5	10,10	10,10	26,601		F11	10,11	10,11	26,611	27,825
G5	10,12	10,12	26,603	29,113	G11	10,10	10,10	26,605	
H5	11,13	11,13	27,856	30,389	H11	10,12	10,12	26,584	29,078
A6	11,11	11,11	27,832						
B6	No data	10,12	26,654	29,180	SUMMARY:				
C6	11,11	11,11	27,862						
D6	11,12	11,12	27,818	29,079	Complete agreement observed among all				
E6	11,12	11,12	27,838	29,049	genotypes.				
F6	8,10	8,10	24,206	26,739					
G6	11,12	11,12	27,843	29,062					
H6	11,11	11,11	27,773						

Exhibit 13. **CDOJ TPOX results compared with ABI 310 and mass spectrometry methods**

Position	ABI 310	Mass Spec	Allele 1 (Da)	Allele 2 (Da)	Position	ABI 310	Mass Spec	Allele 1 (Da)	Allele 2 (Da)
A1	8,10	8,10	17,914	20,431	A7	9,9	5,5	Gas-phase dimer	
B1	6,8	6,8	15,376	17,897	B7	8,8	8,8	17,820	
C1	9,11	9,11	19,137	21,663	C7	10,12	10,12	20,361	22,830
D1	11,12	11,12	21,562	22,801	D7	11,12	6,6	Gas-phase dimer	
E1	9,10	9,10	19,131	20,374	E7	6,8	6,8	15,366	17,850
F1	9,12	9,12	19,106	22,889	F7	10,10	10,10	20,372	
G1	8,8	8,8	17,838		G7	8,9	8,9	17,820	19,070
H1	8,8	8,8	17,825		H7	9,9	9,9	19,098	
A2	8,8	8,8	17,837		A8	9,10	5,5	Gas-phase dimer	
B2	8,8	8,8	17,841		B8	9,9	5,5	Gas-phase dimer	
C2	8,9	8,9	17,860	19,101	C8	8,10	8,10	17,848	20,366
D2	8,9	8,9	17,785	19,049	D8	8,11	8,11	17,843	21,596
E2	8,11	8,11	17,829	21,577	E8	9,9	9,9	19,117	
F2	8,12	8,12	17,834	22,816	F8	9,12	9,12	19,108	22,793
G2	7,10	7,10	16,525	20,257	G8	8,8	8,8	17,905	
H2	8,9	8,9	17,870	19,171	H8	8,10	8,10	17,850	20,366
A3	8,10	8,10	17,844	20,366	A9	8,11	8,11	17,820	21,563
B3	9,10	9,10	19,063	20,320	B9	11,11	11,11	21,575	
C3	9,11	9,11	19,178	21,676	C9	8,8	8,8	17,841	
D3	7,10	7,10	16,626	20,370	D9	10,11	10,11	20,379	21,600
E3	9,11	9,11	19,146	21,665	E9	8,11	8,11	17,829	21,569
F3	8,10	8,10	17,839	20,357	F9	7,8	7,8	16,607	17,837
G3	8,8	8,8	17,853		G9	8,10	8,10	17,904	20,381
H3	6,9	6,9	15,313	19,070	H9	8,11	5,5	Gas-phase dimer	
A4	6,8	6,8	15,359	17,895	A10	9,10	9,10	19,076	20,342
B4	11,11	11,11	21,579		B10	8,11	8,11	17,827	21,569
C4	8,9	8,9	17,865	19,126	C10	8,12	8,12	17,855	22,850
D4	8,9	8,9	17,839	19,088	D10	8,9	8,9	17,846	19,093
E4	6,9	6,9	15,289	19,070	E10	9,10	9,10	19,086	20,346
F4	9,11	5,5	Gas-phase dimer		F10	9,12	9,12	19,065	22,801
G4	8,8	8,8	17,827		G10	11,12	11,12	21,554	22,795
H4	8,8	8,8	17,798		H10	8,11	8,11	17,841	21,584
A5	9,11	9,11	19,126	21,663	A11	6,9	6,9	15,299	19,076
B5	8,11	8,11	17,846	21,586	B11	8,9	8,9	17,851	19,095
C5	10,11	10,11	20,264	21,550	C11	9,11	9,11	19,120	21,592
D5	9,12	9,12	19,137	22,905	D11	8,8	8,8	17,829	
E5	9,10	9,10	19,076	20,337	E11	11,11	11,11	21,569	
F5	9,11	9,11	19,081	21,571	F11	11,11	11,11	21,548	
G5	8,8	8,8	17,831		G11	11,11	11,11	21,579	
H5	6,11	5,5	Gas-phase dimer		H11	8,9	8,9	17,870	19,095
A6	8,11	5,5	Gas-phase dimer		SUMMARY:	Allele	Number	Avg. mass	Std. Dev.
B6	No data	11,11	21,573			6	5	15,343	34.3
C6	8,8	8,8	17,844			7	3	16,586	53.7
D6	8,12	5,10	Gas-phase dimer			8	43	17,846	26.5
E6	8,12	8,12	17,831	22,816		9	31	19,102	33.1
F6	8,11	8,11	17,846	21,600		10	18	20,353	40.7
G6	9,11	9,11	19,070	21,565		11	24	21,590	37.9
H6	9,10	9,10	19,151	20,368		12	10	22,830	39.7

Exhibit 14. **CDOJ TH01 STR results compared with ABI 310 and mass spectrometry methods.** The shaded samples were run with a different primer set and have different masses.

Position	ABI 310	Mass Spec	Allele 1 (Da)	Allele 2 (Da)	Position	ABI 310	Mass Spec	Allele 1 (Da)	Allele 2 (Da)
A1	7,7	7,7	18,854		A7	7,8	7,8	18,877	20,104
B1	7,9.3	7,9.3	17,404	20,854	B7	8,9.3	8,9.3	20,128	22,337
C1	7,8	7,8	18,894	20,108	C7	8,9	8,9	20,193	21,415
D1	6,9	6,9	17,574	21,294	D7	6,9.3	6,9.3	16,167	20,876
E1	7,9.3	7,9.3	18,938	22,387	E7	9.3,9.3	9.3,9.3	22,349	
F1	9,9	9,9	21,415		F7	9,9.3	9,9.3	21,417	22,345
G1	7,9	7,9	18,919	21,408	G7	7,8	7,8	18,911	20,147
H1	5,7	5,7	16,403	18,926	H7	9,9.3	9,9.3	21,400	22,277
A2	6,10	6,10	16,145	21,170	A8	8,8	8,8	20,091	
B2	6,6	6,6	17,662		B8	7,9.3	7,9.3	18,938	22,391
C2	7,8	7,8	18,895	20,115	C8	7,9	7,9	17,436	19,957
D2	7,9	7,9	18,838	21,309	D8	7,7	7,7	18,944	
E2	7,9.3	7,9.3	18,904	22,329	E8	7,7	7,7	17,376	
F2	7,9	7,9	18,831	21,302	F8	7,9	7,9	17,387	19,891
G2	7,7	7,7	17,391		G8	9.3,9.3	9.3,9.3	22,263	
H2	9,9.3	9,9.3	19,871	20,685	H8	7,9.3	7,9.3	18,829	22,267
A3	7,7	7,7	18,856		A9	9,9	9,9	19,907	
B3	7,8	7,8	18,917	20,114	B9	7,7	7,7	18,938	
C3	7,9.3	7,9.3	17,380	20,828	C9	9.3,9.3	9.3,9.3	22,351	
D3	7,9	7,9	17,427	19,944	D9	7,8	7,8	18,845	20,097
E3	7,9.3	7,9.3	18,922	22,364	E9	8,9.3	8,9.3	18,684	20,790
F3	9,9.3	9,9.3	21,377	22,321	F9	6,9.3	6,9.3	17,622	22,287
G3	8,9.3	8,9.3	20,193	22,384	G9	7,9	7,9	17,380	19,885
H3	7,9.3	7,9.3	18,894	22,292	H9	6,7	6,7	17,677	18,926
A4	8,8	8,8	20,126		A10	7,8	7,8	17,389	18,652
B4	6,9.3	6,9.3	17,606	22,294	B10	9,9	9,9	21,377	
C4	6,7	6,7	17,586	18,838	C10	7,9	7,9	17,294	19,812
D4	7,9.3	7,9.3	18,897	22,312	D10	7,8	7,8	17,277	18,547
E4	7,9	7,9	17,401	19,917	E10	6,7	6,7	17,570	18,819
F4	7,9	7,9	18,836	21,334	F10	8,8	8,8	18,640	
G4	8,9.3	8,9.3	20,077	22,258	G10	6,7	6,7	17,644	18,906
H4	7,9.3	7,9.3	18,899	22,304	H10	8,9	8,9	18,643	19,902
A5	7,7	7,7	17,205		A11	6,7	6,7	17,540	18,782
B5	6,9	6,9	16,121	19,874	B11	6,8	6,8	17,682	20,191
C5	8,9.3	8,9.3	20,126	22,360	C11	7,7	7,7	18,927	
D5	6,8	6,8	17,679	20,184	D11	7,7	7,7	17,477	
E5	8,10	8,10	20,155	22,658	E11	6,6	6,6	17,596	
F5	7,9.3	7,9.3	18,870	22,285	F11	7,8	7,8	17,400	18,656
G5	6,7	6,7	17,624	18,906	G11	8,8	8,8	18,710	
H5	7,9	7,9	17,404	19,921	H11	6,7	6,7	17,600	18,895
A6	6,8	6,8	17,648	20,139					
B6	7,7	7,7	18,910		SUMMARY:				
C6	8,9.3	8,9.3	18,626	20,823					
D6	8,8	8,8	20,147		All 88 samples were in agreement.				
E6	8,9.3	8,9.3	19,961	22,143					
F6	8,8	8,8	20,195						
G6	7,9	7,9	18,858	21,383					
H6	8,9	8,9	18,704	19,969					

Exhibit 15. **CDOJ amelogenin results compared with ABI 310 and mass spectrometry methods**

Position	ABI 310	Mass Spec	Allele 1 (Da)	Allele 2 (Da)	Position	ABI 310	Mass Spec	Allele 1 (Da)	Allele 2 (Da)
A1	X,Y	X,Y	25,676	27,566	A7	X,Y	X,Y	25,774	27,687
B1	X,Y	X,Y	25,730	27,596	B7	X,Y	X,Y	25,659	27,555
C1	X,Y	X,Y	25,693	27,667	C7	X,Y	X,Y	25,797	27,695
D1	X,Y	X,Y	25,686	27,607	D7	X,Y	X,Y	25,641	27,533
E1	X,Y	X,Y	25,684	27,765	E7	X,Y	X,Y	25,674	27,583
F1	X,Y	X,Y	25,693	27,592	F7	X,Y	X,Y	25,676	27,585
G1	X,Y	X,Y	25,674	27,574	G7	X,Y	X,Y	25,668	27,568
H1	X,Y	X,Y	25,653	27,551	H7	X,Y	X,Y	25,666	27,564
A2	X,Y	X,Y	25,676	27,574	A8	X,Y	X,Y	25,659	27,555
B2	X,Y	X,Y	25,674	27,579	B8	X,Y	X,Y	25,653	27,544
C2	X,Y	X,Y	25,666	27,559	C8	X,Y	X,Y	25,668	27,566
D2	X,Y	X,Y	25,768	27,626	D8	X,Y	X,Y	25,674	27,566
E2	X,Y	X,Y	25,670	27,568	E8	X,Y	X,Y	25,843	27,763
F2	X,Y	X,Y	25,657	27,544	F8	X,Y	X,Y	25,649	27,581
G2	X,Y	X,Y	25,672	27,646	G8	X,Y	X,Y	25,661	27,561
H2	X,Y	X,Y	25,641	27,523	H8	X,Y	X,Y	25,757	27,585
A3	X,Y	X,Y	25,663	27,559	A9	X,Y	X,Y	25,641	27,531
B3	X,Y	X,Y	25,770	27,581	B9	X,Y	X,Y	25,684	27,579
C3	X,Y	X,Y	25,691	27,635	C9	X,X	X,X	25,645	
D3	X,Y	X,Y	25,722	27,594	D9	X,Y	X,Y	25,653	27,553
E3	X,Y	X,Y	25,749	27,628	E9	X,Y	X,Y	25,674	27,594
F3	X,Y	X,Y	25,682	27,577	F9	X,Y	X,Y	25,720	27,646
G3	X,Y	X,Y	25,645	27,536	G9	X,Y	X,Y	25,766	27,637
H3	X,Y	No data			H9	X,Y	No data		
A4	X,Y	X,Y	25,666	27,553	A10	X,Y	X,Y	25,697	27,618
B4	X,Y	X,Y	25,726	27,637	B10	X,Y	X,Y	25,684	27,590
C4	X,Y	X,Y	25,736	27,583	C10	X,Y	No data		
D4	X,Y	X,Y	25,782	27,670	D10	X,Y	X,Y	25,657	27,564
E4	X,Y	X,Y	25,701	27,789	E10	X,Y	X,Y	25,657	27,549
F4	X,Y	X,Y	25,674	27,585	F10	X,Y	X,Y	25,653	27,551
G4	X,Y	X,Y	25,599	27,572	G10	X,Y	No data		
H4	X,Y	X,Y	25,682	27,579	H10	X,Y	X,Y	25,722	27,654
A5	X,Y	X,Y	25,816	27,719	A11	X,Y	X,Y	25,676	27,572
B5	X,Y	X,Y	25,672	27,570	B11	X,Y	X,Y	25,655	27,549
C5	X,Y	X,Y	25,680	27,583	C11	X,Y	X,Y	25,666	27,577
D5	X,Y	X,Y	25,688	27,661	D11	X,Y	X,Y	25,618	27,464
E5	X,Y	X,Y	25,643	27,523	E11	X,Y	X,Y	25,697	27,676
F5	X,Y	X,Y	25,666	27,572	F11	X,Y	X,Y	25,653	27,557
G5	X,Y	X,Y	25,778	27,672	G11	X,Y	X,Y	25,661	27,553
H5	X,Y	X,Y	25,738	27,568	H11	X,Y	X,Y	25,643	27,540
A6	X,Y	X,Y	25,695	27,587					
B6	X,Y	X,Y	25,649	27,523					
C6	X,Y	X,Y	25,655	27,551					
D6	X,Y	No data							
E6	X,Y	X,Y	25,720	27,605					
F6	X,Y	X,Y	25,674	27,568					
G6	X,Y	X,Y	25,881	27,745					
H6	X,Y	No data							

SUMMARY:		X allele	Y allele
Count		82	81
Average mass		25,690 Da	27,594 Da
Std. dev.		50.4 Da	59.6 Da

Eighty-two samples were in agreement.

"No data" resulted from mass spectrometry in six samples.

Exhibit 16. CDOJ D3S1358 STR results compared with ABI 310 and mass spectrometry methods

Position	ABI 310	Mass Spec	Allele 1 (Da)	Allele 2 (Da)	Position	ABI 310	Mass Spec	Allele 1 (Da)	Allele 2 (Da)
A1	17,17	17,17	28,182		A7	16,17	16,17	27,027	28,112
B1	16,16	16,16	26,967		B7	14,19	14,19	24,305	30,503
C1	15,17	15,17	25,668	28,139	C7	14,14	14,14	24,254	
D1	12,17	12,17	21,820	27,991	D7	15,16	15,16	25,524	26,713
E1	15,15	15,15	25,741		E7	16,16	16,16	26,822	
F1	16,16	16,16	26,762		F7	18,18	18,18	29,491	
G1	14,15	14,15	24,342	25,501	G7	15,16	15,16	25,772	26,830
H1	15,16	15,16	25,802	26,941	H7	16,17	16,17	27,018	28,273
A2	15,17	15,17	25,638	28,134	A8	14,16	14,16	24,331	26,856
B2	16,17	16,17	26,795	27,889	B8	16,17	16,17	26,843	28,056
C2	15,16	15,16	25,661	26,869	C8	15,16	15,16	25,638	26,766
D2	13,16	13,16	23,171	27,091	D8	15,15	15,15	25,724	
E2	14,17	14,17	24,352	28,126	E8	14,15	14,15	24,343	25,477
F2	15,17	15,17	25,549	28,091	F8	15,16	15,16	25,662	26,703
G2	15,17	15,17	25,641	28,119	G8	16,16	16,16	26,907	
H2	14,15	14,15	24,311	25,507	H8	15,15	15,15	25,618	
A3	17,17	17,17	28,184		A9	14,16	14,16	24,323	26,832
B3	18,18	18,18	29,384		B9	15,16	15,16	25,537	26,739
C3	15,15	15,15	25,645		C9	15,15	15,15	25,499	
D3	15,15	15,15	25,645		D9	15,15	15,15	25,522	
E3	16,17	16,17	26,967	27,945	E9	14,15	14,15	24,440	25,459
F3	15,16	15,16	25,573	26,790	F9	16,17	16,17	26,745	28,021
G3	17,18	17,18	28,025	29,302	G9	17,18	17,18	28,372	29,315
H3	14,15	14,15	24,360	25,530	H9	15,16	15,16	25,761	26,777
A4	17,17	17,17	28,001		A10	16,16	16,16	26,913	
B4	15,16	15,16	25,518	26,811	B10	15,18	15,18	25,539	29,347
C4	15,16	15,16	25,705	26,722	C10	15,16	15,16	25,477	26,741
D4	15,17	15,17	25,589	28,110	D10	14,15	14,15	24,214	25,341
E4	16,17	16,17	27,035	28,023	E10	14,15	14,15	24,252	25,507
F4	14,16	14,16	24,374	26,886	F10	16,17	16,17	26,871	28,139
G4	14,17	14,17	24,342	28,115	G10	16,17	16,17	27,033	27,934
H4	15,18	15,18	25,655	29,378	H10	15,16	15,16	25,716	26,715
A5	15,16	15,16	25,727	27,110	A11	15,18	15,18	25,545	29,293
B5	16,17	16,17	26,988	28,300	B11	15,15.2	15,15.2	25,734	26,379
C5	15,15	15,15	25,684		C11	16,17	16,17	26,937	27,845
D5	14,16	14,16	24,520	26,941	D11	15,16	15,16	25,544	26,672
E5	15,17	15,17	25,678	28,132	E11	16,16	16,16	26,899	
F5	16,16	16,16	26,813		F11	15,16	15,16	25,706	26,678
G5	9,17	9,17	18,092	28,285	G11	16,16	16,16	26,841	
H5	17,17	17,17	27,893		H11	16,17	16,17	26,881	28,246
A6	16,17	16,17	26,999	28,178					
B6	15,17	15,17	25,645	28,182					
C6	15,16	15,16	25,681	26,710		SUMMARY:			
D6	15,15	15,15	25,833						
E6	14,16	14,16	24,301	26,816		All 88 samples were in agreement.			
F6	14,14	14,14	24,437						
G6	14,15	14,15	24,425	25,524		ABI 310 CE electropherogram of boxed data is shown in exhibit 74.			
H6	15,15	15,15	25,523						

Exhibit 17. **CDOJ D16S539 STR results compared with ABI 310 and mass spectrometry methods**

Position	ABI 310	Mass Spec	Allele 1 (Da)	Allele 2 (Da)	Position	ABI 310	Mass Spec	Allele 1 (Da)	Allele 2 (Da)
A1	11,12	11,12	26,862	28,095	A7	11,11	11,11	26,873	
B1	9,14	9,14	24,421	30,665	B7	9,10	9,10	24,338	25,595
C1	9,11	9,11	24,360	26,888	C7	11,13	11,13	26,894	29,471
D1	10,11	10,11	25,622	26,831	D7	11,11	11,11	26,879	
E1	11,13	11,13	26,903	29,367	E7	9,12	9,12	24,425	28,200
F1	10,11	10,11	25,543	26,830	F7	9,12	9,12	24,368	28,160
G1	11,12	11,12	26,852	28,070	G7	11,12	11,12	26,997	28,145
H1	8,9	8,9	23,157	24,360	H7	12,12	12,12	28,104	
A2	9,11	9,11	24,342	26,805	A8	9,13	9,13	24,360	29,362
B2	12,13	12,13	28,191	29,323	B8	9,9	9,9	24,394	
C2	11,11	11,11	26,907		C8	9,13	9,13	24,352	29,369
D2	13,13	13,13	29,445		D8	11,11	11,11	26,933	
E2	12,12	12,12	28,128		E8	8,12	8,12	23,332	28,486
F2	12,13	12,13	28,202	29,436	F8	11,12	11,12	26,937	28,128
G2	10,11	10,11	25,638	26,869	G8	9,13	9,13	24,336	29,351
H2	10,11	10,11	25,576	26,864	H8	11,14	11,14	26,864	30,573
A3	9,12	9,12	24,340	28,106	A9	11,13	11,13	26,869	29,378
B3	9,12	9,12	24,356	28,117	B9	11,12	11,12	26,888	28,078
C3	12,13	12,13	28,141	29,483	C9	12,12	12,12	28,106	
D3	9,9	9,9	24,417		D9	11,13	11,13	26,849	29,342
E3	11,13	11,13	27,014	29,603	E9	11,13	11,13	26,899	29,396
F3	11,11	11,11	26,899		F9	12,15	12,15	28,115	31,845
G3	9,11	9,11	24,376	26,931	G9	11,13	11,13	26,886	29,389
H3	8,10	8,10	23,161	25,649	H9	11,13	11,13	26,888	29,387
A4	9,11	9,11	24,427	26,935	A10	9,11	9,11	24,423	26,920
B4	9,12	9,12	24,480	28,248	B10	9,12	9,12	24,449	28,126
C4	13,13	13,13	29,318		C10	12,13	12,13	28,302	29,498
D4	11,12	No data			D10	8,9	8,9	23,072	24,317
E4	13,13	13,13	29,389		E10	9,13	9,13	24,376	29,402
F4	9,11	9,11	24,374	26,894	F10	11,13	11,13	26,856	29,436
G4	13,14	No data			G10	9,9	9,9	24,358	
H4	11,13	11,13	26,837	29,336	H10	10,13	10,13	25,653	29,441
A5	12,12	12,12	28,117		A11	9,12	9,12	24,427	28,128
B5	8,9	8,9	23,181	24,417	B11	11,12	11,12	26,862	28,043
C5	10,11	10,11	25,670	27,007	C11	11,12	No data		
D5	12,12	12,12	28,265		D11	11,11	11,11	26,890	
E5	9,11	9,11	24,482	26,984	E11	11,13	11,13	26,928	29,619
F5	9,11	9,11	24,329	26,854	F11	9,10	No data		
G5	10,12	10,12	25,688	28,222	G11	10,12	No data		
H5	11,12	11,12	26,918	28,041	H11	11,11	11,11	26,894	
A6	9,11	9,11	24,372	26,888					
B6	No data	9,12	24,405	28,208	SUMMARY:				
C6	9,13	9,13	24,380	29,398	Eighty-one samples were in agreement.				
D6	9,13	9,13	24,465	29,416	"No data" resulted from mass spectrometry in five samples.				
E6	12,12	12,12	28,170		"No data" resulted from ABI 310 in one sample.				
F6	13,13	13,13	29,329		Gas-phase trimer caused an error in one sample.				
G6	11,13	11,13	26,867	29,371					
H6	11,11	9,9	Gas-phase trimer						

Exhibit 18. CDOJ D7S820 STR results compared with ABI 310 and mass spectrometry methods

Position	ABI 310	Mass Spec	Allele 1 (Da)	Allele 2 (Da)	Position	ABI 310	Mass Spec	Allele 1 (Da)	Allele 2 (Da)
A1	11,13	11,13	20,322	22,807	A7	10,12	10,12	19,097	21,621
B1	10,11	10,11	19,081	20,333	B7	11,12	11,12	20,545	21,592
C1	8,11	8,11	16,641	20,379	C7	8,10	10,10	19,360	
D1	9,10	9,10	17,857	19,131	D7	11,12	11,12	20,383	21,646
E1	8,10	10,10	19,072		E7	9,12	9,12	17,841	21,583
F1	10,12	10,12	19,088	21,594	F7	8,9	8,8	16,631	
G1	8,10	8,10	16,643	19,174	G7	11,12	11,12	20,374	21,575
H1	8,10	8,10	16,616	19,088	H7	11,12	11,12	20,366	21,581
A2	8,11	8,11	16,646	20,448	A8	10,12	10,12	19,189	21,703
B2	10,11	10,11	19,169	20,381	B8	10,11	10,11	19,265	20,400
C2	8,8	8,8	16,534		C8	10,10	10,10	19,095	
D2	10,10	10,10	19,189		D8	9,10	9,10	17,848	19,093
E2	8,8	8,8	16,641		E8	8,10	10,10	19,169	
F2	8,8	8,8	16,616		F8	8,9	No data	*Double null alleles?*	
G2	8,10	10,10	19,101		G8	9,12	9,12	17,949	21,745
H2	8,9	8,9	16,550	17,834	H8	8,10	10,10	19,187	
A3	11,12	11,12	20,372	21,569	A9	8,11	11,11	20,571	
B3	10,11	10,11	19,104	20,366	B9	8,8	8,8	16,534	
C3	8,12	8,12	16,532	21,548	C9	10,10	10,10	19,169	
D3	12,13	12,13	21,560	22,789	D9	7,8	7,8	15,383	16,638
E3	10,12	10,12	19,187	21,695	E9	9,10	9,10	17,844	19,090
F3	10,11	10,11	19,198	20,381	F9	10,12	10,12	19,395	21,866
G3	8,11	11,11	20,374		G9	8,10	8,10	16,651	19,201
H3	8,10	10,10	19,097		H9	8,11	8,11	16,826	20,363
A4	9,9	9,9	17,848	17,848	A10	12,13	12,13	21,586	22,820
B4	11,11	11,11	20,528		B10	11,12	11,12	20,387	21,586
C4	8,12	8,12	16,639	21,667	C10	8,10	10,10	19,330	
D4	8,9	8,9	16,636	17,860	D10	8,11	8,11	16,554	20,342
E4	9,11	11,11	20,392		E10	10,11	10,11	19,086	20,348
F4	10,11	10,11	19,185	20,396	F10	9,10	9,10	17,851	19,097
G4	10,10	10,10	19,101		G10	9,10	10,10	19,398	
H4	10,12	12,12	21,579		H10	10,11	10,11	19,382	20,377
A5	9,10	9,10	17,836	19,083	A11	10,10	10,10	19,254	
B5	11,12	11,12	20,631	21,678	B11	10,11	10,11	19,093	20,353
C5	11,12	11,12	20,363	21,573	C11	9,10	9,10	17,886	19,187
D5	10,10	10,10	19,088		D11	8,11	8,11	16,643	20,519
E5	9,11	9,11	17,851	20,372	E11	10,10	10,10	19,252	
F5	8,11	8,11	16,634	20,377	F11	9,12	9,12	17,831	21,569
G5	11,12	11,12	20,590	21,787	G11	10,12	10,12	19,142	21,653
H5	10,10	10,10	19,322		H11	10,12	10,12	19,160	21,883
A6	8,10	8,10	16,556	19,074					
B6	8,9	8,8	16,643						
C6	8,11	11,11	20,374						
D6	8,10	8,10	16,614	19,090					
E6	8,9	9,9	17,851						
F6	8,8	8,8	16,831						
G6	8,10	10,10	19,155						
H6	9,9	9,9	18,169						

SUMMARY:

Seventy samples were in agreement.

"No data" resulted from mass spectrometry (null alleles?) in one sample.

Seventeen samples disagreed because of null alleles.

Exhibit 19. **CDOJ FGA STR results compared with ABI 310 and mass spectrometry methods**

Position	ABI 310	Mass Spec	Allele 1 (Da)	Allele 2 (Da)	Position	ABI 310	Mass Spec	Allele 1 (Da)	Allele 2 (Da)
A1	20,21	No data			A7	23,24	24,24	40,518	Not resolved
B1	20,23	20,23	35,987	39,547	B7	22,25	22,25	38,290	41,811
C1	23,25	23,25	39,390	41,917	C7	24,25	24,24	41,079	Not resolved
D1	19,23	19,23	34,858	39,547	D7	22,22	21,21	37,930	
E1	23,24	23,23	40,317	Not resolved	E7	24,27	24,27	40,680	44,130
F1	18.2,26	18.2,26	34,066	43,221	F7	22,22	22,22	38,295	
G1	21,21	21,21	37,011		G7	22,23	22,22	38,582	Not resolved
H1	No data	23,23	39,555		H7	21,26	21,26	37,148	43,116
A2	22,23	23,23	39,446	Not resolved	A8	OL?,24	16,24	31,415	40,688
B2	21,22	21,21	37,570	Not resolved	B8	22,24	22,24	38,226	40,874
C2	21,25	21,25	37,146	41944	C8	19.2,25	19.2,25	35,193	41,718
D2	19.19.2	19,19	34,699	Not resolved	D8	24,25	24,24	41,184	Not resolved
E2	23,27	23,26	39,310	43958	E8	18.2,24	18,23	33,960	40,247
F2	19,25	19,25	34,667	41575	F8	23,25	23,25	39,482	41,914
G2	22,22	22,22	38,554		G8	25,25	24,24	41,668	
H2	22,25	22,25	38,244	41,861	H8	21,23	21,23	37,113	39632
A3	22,24	22,24	38,110	40,604	A9	19,25	19,25	34,631	41,827
B3	30.2, OL?	30,30	48,114		B9	20,24	20,24	35,888	40,688
C3	21,24	21,24	37,158	40,777	C9	20,24	19,23	35,667	40,312
D3	23,23	23,23	39,475		D9	21,24	21,24	36,973	40,573
E3	19,23	19,23	34,769	39,547	E9	21,25	21,24	36,921	41,549
F3	23,24	23,23	39,852	Not resolved	F9	25,26	25,25	42,218	Not resolved
G3	22,23	22,22	38,582	Not resolved	G9	20,21	21,21	36,901	Not resolved
H3	18.2,23	18.2,23	34,138	39,423	H9	18.2,25	18.2,25	34,111	41,774
A4	22,23	22,22	39,056	Not resolved	A10	24,28	24,28	40,706	45,467
B4	19,26	19,26	34,578	42761	B10	21,24	21,24	37,018	40,636
C4	24,24	24,24	40,649		C10	18,23	18,23	33,560	39,490
D4	26,27	No data			D10	22,23	22,22	38,537	Not resolved
E4	22,22	22,22	38,442		E10	18,23	18,23	33,292	39,289
F4	22,23	22,22	39,081	Not resolved	F10	19,22	19,22	34,829	38,402
G4	23,24	23,23	39,741	Not resolved	G10	22,24	22,24	38,216	40,581
H4	23,23	23,23	39,562		H10	20,26	20,26	36,127	43,407
A5	20,22	20,22	35,856	38,364	A11	24,26	24,26	40,659	43,043
B5	23,23	22,22	39,017		B11	23,28	23,28	39,493	45,376
C5	22,23	22,22	38,659	Not resolved	C11	24,25	24,24	40,869	Not resolved
D5	23,26	22,25	39,233	42,879	D11	22,25	22,25	38,361	41,989
E5	22,23	22,22	38,554	Not resolved	E11	25,26	25,25	42,333	Not resolved
F5	19,25	19,25	34,795	41,994	F11	22,24	22,24	38,209	40,560
G5	21,24	21,24	37,366	40,853	G11	23,25	23,25	39,511	41,909
H5	21,22	21,21	37,565	Not resolved	H11	23,26	23,26	39,408	42,965
A6	22,24	22,24	38,155	40,570					
B6	24,27	24,27	40,612	44,288					
C6	21,25	21,25	37,136	41,997					
D6	20,23	20,23	35,972	39,477					
E6	22,23	22,22	38,644	Not resolved					
F6	25,28	25,27	41,737	45,299					
G6	18.2,24	18.2,24	34,387	40,869					
H6	21,23	21,23	37,113	39,534					

SUMMARY:

Fifty-three samples were in agreement.

"No data" resulted from mass spectrometry in two samples.

"No data" resulted from ABI 310 in one sample.

One sample was too large for detection.

Nine samples resulted in wrong calls from poor calibration.

Twenty-two samples resulted in wrong calls from poor resolution.

SCOPE AND METHODOLOGY

Both the STR and the SNP geno-typing assays used in this project involve the same fundamental (proprietary) sample preparation chemistry. This chemistry was important for salt reduction/removal prior to the mass spectrometric analysis and was automated on a 96-tip robotic workstation. A biotinylated, cleavable oligonucleotide was used as a primer in each assay and was incorporated through standard DNA amplification (i.e., PCR) methodologies into the final product, which was measured in the mass spectrometer. This process was covered by U.S. Patent 5,700,642, which was issued in December 1997, and is described in more detail in U.S. Patent 6,090,588 (Butler et al., 2000). The STR assay is schematically illustrated in exhibit 20 and involves a PCR amplification step where one of the primers is replaced by the GeneTrace cleavable primer. The biotinylated PCR product was then captured on streptavidin-coated magnetic beads for post-PCR sample cleanup and salt removal followed by mass spectrometry analysis.

The biology portion of the SNP assay, on the other hand, involves a three-step process: (1) PCR amplification, (2) phosphatase removal of nucleotides, and (3) primer extension, using the GeneTrace cleavable primer, with dideoxynucleotides for single-base addition of the nucleotide(s) complementary to the one(s) at the SNP site (Li et al., 1999). The SNP assay is illustrated in exhibit 21.

Simultaneous analysis of multiple SNP markers (i.e., multiplexing) is possible by simply putting the cleavage sites at different positions in the various primers so they do not overlap on a mass scale. Also important to both genotyping assays is proprietary calling software that was developed (and evolved) during the course of this work. A number of STR and SNP markers were developed and tested with a variety of human DNA samples as part of this project to demonstrate the feasibility of this mass spectrometry approach.

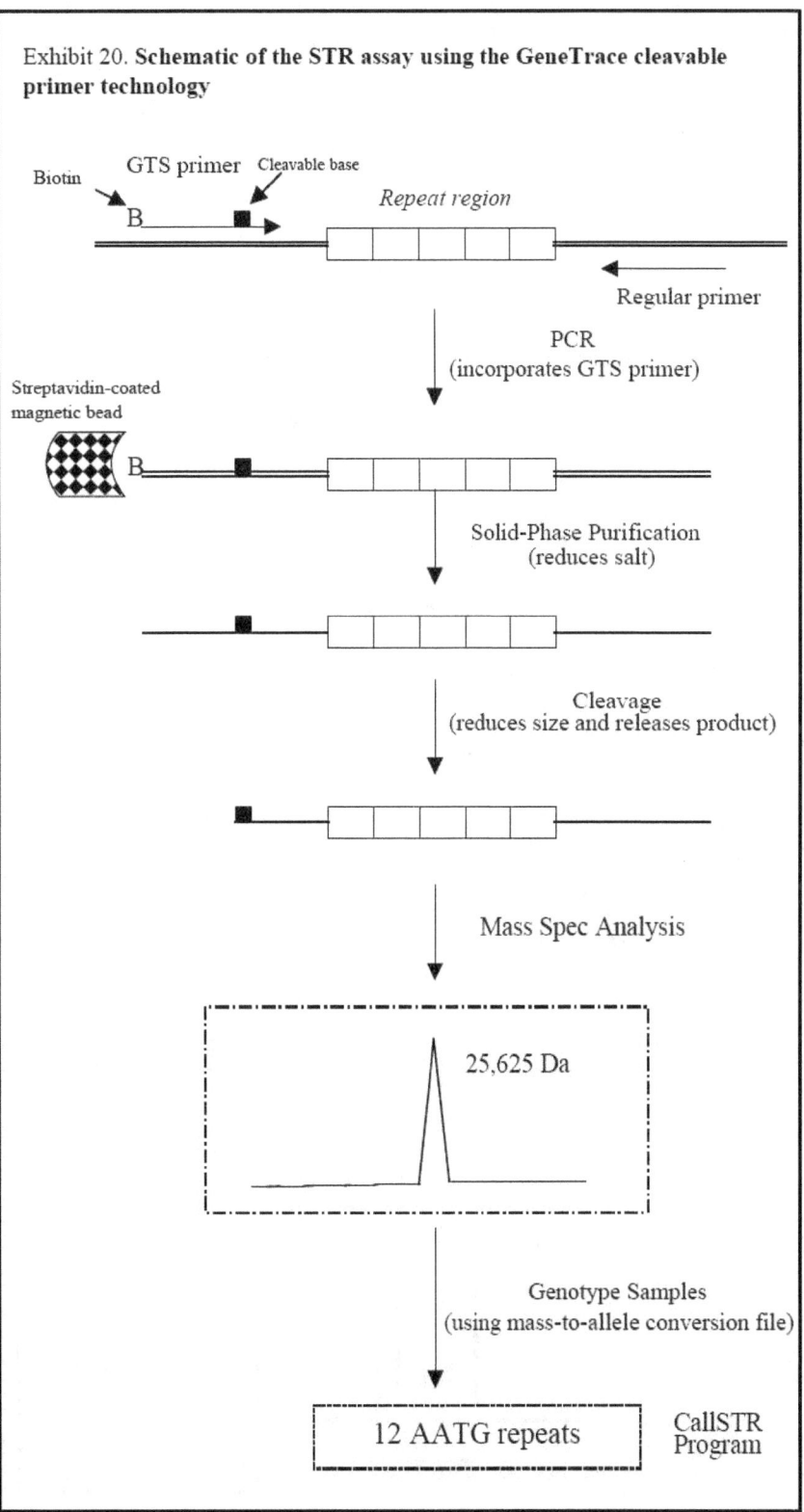

Exhibit 20. **Schematic of the STR assay using the GeneTrace cleavable primer technology**

Assay Development and Primer Testing

Primer design

Primers were initially designed for each STR locus using Gene Runner software (Hastings Software, Inc., Hastings, NY) and then more recently with Primer 3 version 0.2 from the World Wide Web (*http://www-genome.wi.mit.edu/cgi-bin/primer/primer3_www.cgi*) (Rozen and Skaletsky, 1998). Multiplex PCR primers for the multiplex SNP work were designed with a UNIX version of Primer 3 (release 0.6) adapted at GeneTrace to utilize a mispriming library and Perl scripts for input of sequences and export of primer information.

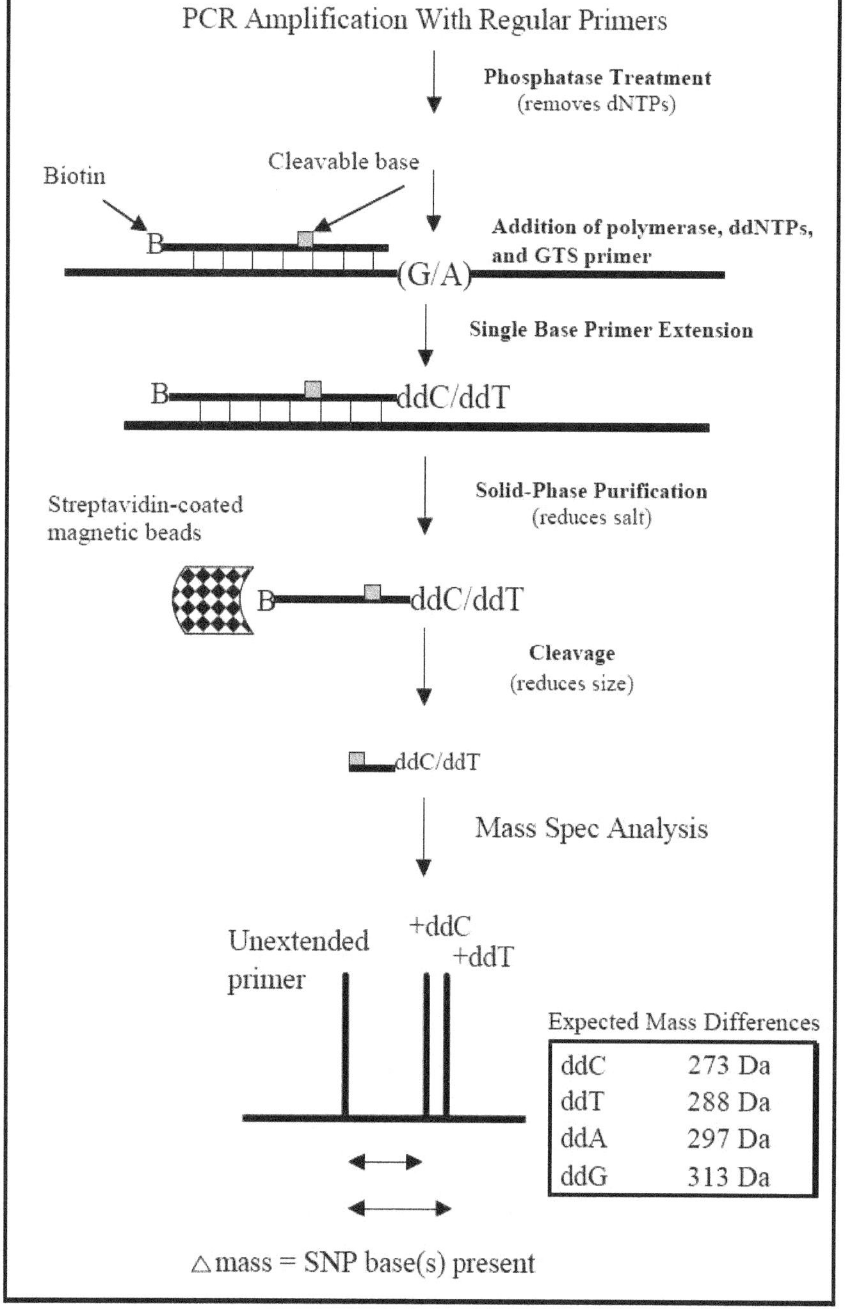

DNA sequence information was obtained from GenBank (*http://www.ncbi.nlm.nih.gov*) and STRBase (*http://www.cstl.nist.gov/biotech/strbase*) for the STR loci and mtDNA and from Dr. Peter Underhill of Stanford University for the Y-chromosome SNPs. These sequences served as the reference sequence for primer design and, in the case of STRs, the calibrating mass for the genotyping software (see below). When possible, primers were placed close to the repeat region to make the PCR product size ranges under 120 bp to improve the sensitivity and resolution in the mass spectrometer (exhibit 2). Previously published primers were used in the case of amelogenin (Sullivan et al., 1993), D3S1358 (Li et al., 1993), CD4 (Hammond et al., 1994), and VWA (Fregeau and Fourney, 1993) because their PCR product sizes were analyzable in the mass spectrometer or the amplicons could be reduced in size following the PCR step (see below). Later, D3S1358 experiments were performed with primers that produced smaller products after sequence information became available for that particular STR locus (exhibit 22).

Primer synthesis

Unmodified primers were purchased from Biosource/Keystone (Foster City, CA) or Operon Technologies (Alameda, CA) or synthesized in-house using standard solid-phase phosphoramidite chemistry. The GeneTrace cleavable primers were synthesized in-house using a proprietary phosphoramidite that was incorporated near the 3' end of the oligonucleotide along with a biotin attached at the 5' end. Primers were quality control tested via mass spectrometry prior

Exhibit 21. **Schematic of the SNP assay using the GeneTrace cleavable primer technology**

PCR Amplification With Regular Primers

Phosphatase Treatment (removes dNTPs)

Biotin

Cleavable base

B

(G/A)

Addition of polymerase, ddNTPs, and GTS primer

Single Base Primer Extension

B ddC/ddT

Streptavidin-coated magnetic beads

Solid-Phase Purification (reduces salt)

B ddC/ddT

Cleavage (reduces size)

ddC/ddT

Mass Spec Analysis

Unextended primer

+ddC +ddT

Expected Mass Differences

ddC	273 Da
ddT	288 Da
ddA	297 Da
ddG	313 Da

△mass = SNP base(s) present

Exhibit 22. **D3S1358 sequence with PCR primer locations.** This STR sequence was not publicly available in GenBank at the start of this project but was obtained as part of this work. The forward primer is shown in blue and the reverse primer in red, with the GATA repeat region shaded grey. Section (A) shows the primer locations compared with the repeat region for a primer pair originally published by Li and colleagues (1993). Section (B) shows the primers designed as part of this project and their positions relative to the repeat region. The overall size of PCR products was reduced by 27 bp (more than 8,000 Da) compared with an amplicon generated from the previously published primer set.

(A) Previously Published Primers: PCR product size = 127 bp for 15 repeats

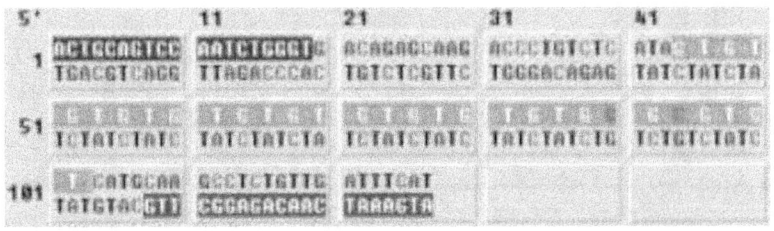

(B) Newly Designed Primers: PCR product size = 100 bp for 15 repeats

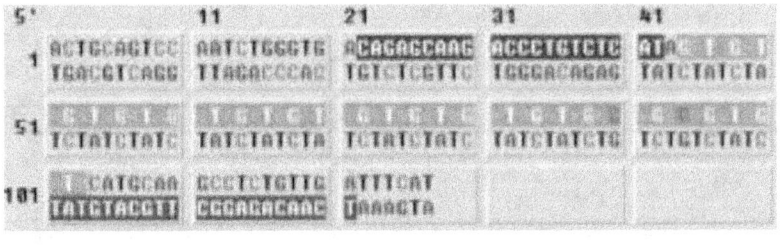

to further testing to confirm proper synthesis and to determine the presence or absence of failure products. Synthesis failure products (i.e., n-1, n-2, etc.) can especially interfere with multiplex SNP analysis. The cleavable base is stable during primer synthesis and PCR amplification. Comparisons of regular primers with cleavable primers containing the same base sequence showed no significant difference, indicating that the primer annealing is not compromised by the cleavable base.

Methods for STR product size reduction

It was discovered early in the study that the PCR primer opposite the biotinylated cleavable primer could be moved into the repeat region as much as two full repeat units to reduce the overall size without severely compromising the PCR reaction. For the cleavable primer, the cleavable base was typically placed in the second or third position from the 3' end of the primer in order to remove as much of the modified primer as possible. Thus, the cleavage step reduces the overall PCR product size by the length of the cleavable primer minus two or three nucleotides. Typically, this size reduction is approximately 20 bases.

The portion of the DNA product on the other side of the repeat region from the cleavable primer was removed in one of two possible ways: using a restriction enzyme (Monforte et al., 1999) or performing a nested linear amplification with a terminating nucleotide (Braun

et al., 1997a and 1997b), such as dideoxynucleotide (ddN). These methods work only for particular situations (see Results and Discussion). Almost all singleplex STR work was performed without either of these product size reduction methods. However, these size reduction methods played a role in the multiplex STR work.

Multiplex design

STR multiplexes were designed by construction of virtual allelic ladders or "mass simuplexes" that involved the predicted mass of all known alleles for a particular locus. STR markers were then interleaved based on mass with all alleles between loci being distinguishable (exhibit 4). STR multiplexes work best if alleles are below 20,000–25,000 Daltons (Da) in mass due to the improved sensitivity and resolution that is obtainable in the mass spectrometer. As previously described in the section on size reduction, a restriction enzyme or a ddN terminator may be used to shorten the STR allele sizes. For multiplex design, locating a restriction enzyme with cut sites common to all STR loci involved in the multiplex complicates the design process and limits the choice of possible marker combinations. The use of a common dideoxynucleotide terminator is much easier. For example, with the STR loci CSF1PO, TPOX, and TH01, a multiplex was developed using a dideoxycytosine (ddC) terminator and primer extension along the AATG strand (exhibits 4 and 5).

SNP multiplexes were designed by calculating possible postcleavage primer and extension product masses. Multiply charged ions were abundant in the mass range of 1,500–7,000 Da in SNP multiplex analyses, which were avoided for the most part by calculating interfering doubly charged and triply charged ions. The cleavage sites for candidate multiplex SNP primers were chosen for the least amount of overlap between singly and multiply charged ions (exhibits 23).

Human DNA samples used

Human genomic DNA samples representing several ethnic groups (African-American, European, and Asian) were purchased from Bios Laboratories (New Haven, CT) for the initial studies. K562 cell line DNA (Promega) was used as a control sample since the genotypes for this cell line were reported in most of the STR loci (*GenePrint*™ STR Systems Technical Manual, 1995).

Allelic ladders were reamplified from a 1:1000 dilution of each of the allelic ladders supplied in fluorescent STR kits from ABI using the PCR conditions listed below and the primers shown in exhibit 24. The ABI kits included allelic

Exhibit 23. **Expected mass-to-charge ratios for various ions in the mtDNA 10-plex assay**

Name	Primer Mass	Singly Charged Ions				Primer Mass	Doubly Charged Ions				Primer Mass	Triply Charged Ions			
		ddC	ddT	ddA	ddG		ddC	ddT	ddA	ddG		ddC	ddT	ddA	ddG
MT5	1,580			1,877	1,893	790			939	947	527			626	631
MT8'	1,790	2,063	2,078			895	1,032	1,039			597	688	693		
MT10	2,785	3,058	3,073			1,393	1,529	1,537			928	1,019	1,024		
MT3'	3,179			3,476	3,492	1,590			1,738	1,746	1,060			1,159	1,164
MT9	3,740			4,037	4,053	1,870			2,019	2,027	1,247			1,346	1,351
MT6	4,355			4,652	4,668	2,178			2,326	2,334	1,452			1,551	1,556
MT2g	4,957	5,230	5,245			2,479	2,615	2,623			1,652	1,743	1,748		
MT1	5,375			5,672	5,688	2,688			2,836	2,844	1,792			1,891	1,896
MT7	5,891	6,164	6,179			2,946	3,082	3,090			1,964	2,055	2,060		
MT4e	6,500	6,773	6,788			3,250	3,387	3,394			2,167	2,258	2,263		

Exhibit 24. **Primer sequences designed for STR markers tested by mass spectrometry.** For the PCR product size produced with these primers, see exhibit 2. Note: "b" is listed for biotin; parentheses ( ) indicate the cleavable base.

STR Locus	Primer Sequences for GTS Mass Spec Analysis	Primer Name	STR Locus	Primer Sequences for GTS Mass Spec Analysis	Primer Name
Amelogenin	5'-b-CCCTGGGCTCTGTAAAGAATAG(T)G-3' 5'-ATCAGAGCTTAAACTGGGAAGCTG-3'	AMEL-F AMEL-R	D21S11	5'-b-CCCAAGTGAATTGCCTTC(T)A-3' 5'-GTAGATAGACTGGATAGATAGACGATAGA-3'	D21-F D21-R
CD4	5'-b-TTGGAGTCGCAAGCTGAAC(T)AGC-3' 5'-GCCTGAGTGACAGAGTGAGAACC-3'	CD4-F CD4-R	F13A1	5'-b-CAGAGCAAGACTTCATC(T)G-3' 5'-TCATTTTAGTGCATGTTC-3'	F13A1-F F13A1-R
CSF1PO	5'-ACAGTAACTGCCTTCATAGA(T)AG-3' 5'-b-GTGTCAGACCCTGTTCTAAGTA-3'	CSF-F3 CSF-R3	FES/FPS	5'-b-TTAGGAGACAAGGATAGCAGT(T)C-3' 5'-GCGAAAGAATGAGACTACATCT-3'	FES-F2 FES-R2
D3S1358	5'-b-CAGAGCAAGACCCTGTC(T)CAT-3' 5'-TCAACAGAGGCTTGCATGTAT-3'	D3-F2 D3-R2	FGA	5'-b-AAAATTAGGCATATTTACAAGCTAG(T)T-3' 5'-TCTGTAATTGCCAGCAAAAAAGAAA-3'	FGA-F FGA-R
D5S818	5'-b-CTCTTTGGTATCCTTATGTAATA(T)T-3' 5'-ATCTGTATCCTTATTTATACCTCTATCTA-3'	D5-F D5-R	HPRTB	5'-b-GTCTCCATCTTTGTCTCTATCTCTATC(T)G-3' 5'-GAGAAGGGCATGAATTTGCTTT-3'	HPRTB-F HPRTB-R
D7S820	5'-b-TGTCATAGTTTAGAACGAAC(T)AAC-3' 5'-AAAAACTATCAATCTGTCTATCTATC-3'	D7-F D7-R	LPL	5'-b-CTGACCAAGGATAGTGGGATA(T)AG-3' 5'-GGTAACTGAGCGAGACTGTGTCT-3'	LPL-F LPL-R
D8S1179	5'-b-TTTGTATTTCATGTGTACATTCGTA(T)C-3' 5'-ACCTATCCTGTAGATTATTTTCACTGTG-3'	D8-F D8-R	TH01	5'-CCTGTTCCTCCCTTATTTCCC-3' 5'-b-GGGAACACAGACTCCATGG(T)G-3'	TH01-F TH01-R
D13S317	5'-b-CCCATCTAACGCCTATCTGTA(T)T-3' 5'-GCCCAAAAAGACAGACAGAAAG -3'	D13-F D13-R2	TPOX	5'-b-CTTAGGGAACCCTCACTGAA(T)G-3' 5'-b-GTCCTTGTCAGCGTTTATTTGC-3'	TPOX-F TPOX-R
D16S539	5'-b-ATACAGACAGACAGACAGG(T)G-3' 5'-GCATGTATCTATCATCCATCTCT-3'	D16-F4 D16-R4	VWA	5'-b-CCCTAGTGGATGATAAGAATAATCAGTATG-3' 5'-b-GGACAGATGATAAATACATAGGATGGA(T)GG-3'	VWA-F VWA-R
D18S51	5'-TGAGTGACAAATTGAGACCTT-3' 5'-b-GTCTTACAATAACAGTTGCTACTA(T)T-3'	D18-F D18-R			

ladders for the following STR loci: AmpF1STR® Green I (CSF1PO, TPOX, TH01, amelogenin), AmpF1STR® Blue (D3S1358, VWA, FGA), AmpF1STR® Green II (amelogenin, D8S1179, D21S11, D18S51), AmpF1STR® Yellow (D5S818, D13S317, D7S820), and AmpF1STR® COfiler™ (amelogenin, TH01, TPOX, CSF1PO, D3S1358, D16S539, D7S820).

While most PCR amplifications were performed with quantitated genomic DNA in liquid form, a few were tested with blood-stained FTA™ paper (Life Technologies, Rockville, MD). Sample punches were removed from the dried FTA™ paper card with a 1.2 mm Harris MICRO-PUNCH™ (Life Technologies). The recommended washing protocol of 200 µL was reduced to 25 or 50 µL in order to reduce reagent costs and to work with volumes that are compatible with 96- or 384-well sample plates. The number of washes was kept the same as recommended by the manufacturer, but deionized water was used instead of a 10 mM Tris-EDTA solution.

Two studies were performed with larger numbers of DNA samples. In collaboration with Dr. Steve Lee and Dr. John Tonkyn from CDOJ's DNA research laboratory, CDOJ provided a plate of 88 samples, which was used repeatedly for multiple STR markers. These anonymous samples had been previously genotyped by CDOJ using ABI's AmpF1STR® Profiler™ kit, which consists of AmpF1STR® Blue, Green I, and Yellow markers and amplifies 9 STRs and the sex-typing marker amelogenin. STR allelic ladders were also provided by CDOJ and were used to illustrate that the common alleles for each STR locus could be detected with GeneTrace primer sets. Researchers retyped the samples using the AmpF1STR® COfiler™ fluorescent STR kit, which contains 6 STRs and amelogenin (5 of the 6 STR loci overlap with Profiler loci), thereby providing

a further validation of each sample's true genotype. More recently, a set of 92 human DNA templates containing 3 different Centre d'Etude du Polymorphisme Humain (CEPH) families (exhibit 25) and 44 unrelated individuals from the NIH Polymorphism Discovery Resource were examined (Collins et al., 1998). These samples were typed on the ABI 310 Genetic Analyzer using both the AmpF1STR® Profiler Plus™ and AmpF1STR® COfiler™ kits so that all 13 CODIS STRs were covered.

PCR reaction

To speed the development of new STR markers, researchers worked toward the development of universal PCR conditions, in terms of both thermal cycling parameters and reagents used. Since almost all amplifications were singleplex PCRs, development effort was much simpler than multiplex PCR development. Generally, all PCR reactions were performed in 20 µL volumes with 20 pmol (1 µM) both forward and reverse primers and a PCR reaction mix containing everything else. The

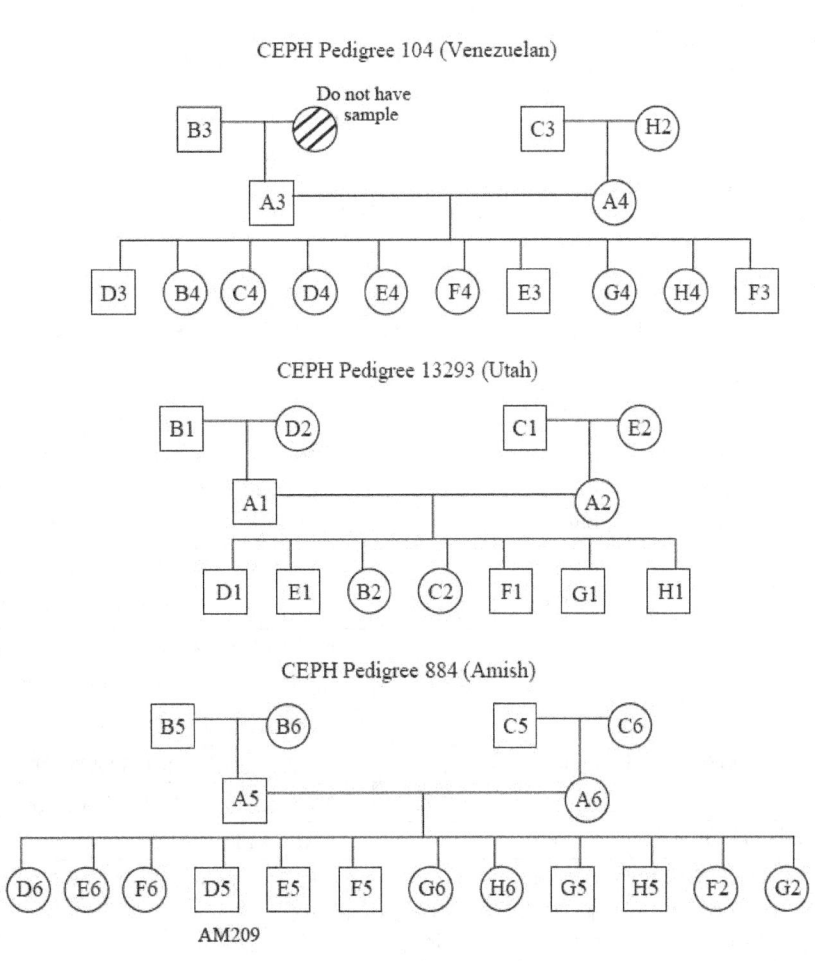

Exhibit 25. **CEPH family pedigrees for samples examined.** The 96 well positions are indicated for each sample. These samples were typed at 13 STRs and amelogenin using Profiler Plus™ and COfiler™ fluorescent STR kits. The samples were also typed at 8 STRs and amelogenin using the GeneTrace mass spectrometry primers.

CEPH Pedigree 104 (Venezuelan)

CEPH Pedigree 13293 (Utah)

CEPH Pedigree 884 (Amish)

early PCR reaction mix contained 1 U Taq polymerase (Promega); 1X STR buffer with deoxynucleotide triphosphates (dNTPs) (Promega); and typically 5, 10, or 25 ng of human genomic DNA. Later in the study, a PCR mix containing 200 µM dNTPs, 50 mM KCl, 10 mM Tris-HCl, 5% glycerol, and 2 mM MgCl$_2$ was used. Typically, a locus-specific master mix was prepared by the addition of 12.8 µL of PCR mix times the number of samples (+ ~10% overfill) with 0.2 µL AmpliTaq Gold™ DNA polymerase (ABI) and the appropriate volume and quantity of forward and reverse primers to bring them to a concentration of 1 µM in each reaction. PCR reactions in a 96- or 384-well format were set up manually with an 8-channel pipettor or robotically with a Hamilton 16-tip robot.

Thermal cycling was performed in 96- or 384-well MJ Research DNA Engine (MJ Research, Watertown, MA) or 96 or dual block 384 PE9700 (ABI) thermal cyclers.

Initial thermal cycling conditions with Taq polymerase (Promega) were as follows:

94 °C for 2 min
35 cycles:
    94 °C for 30 sec
    50, 55, or 60 °C for 30 sec
    72 °C for 30 sec
72 °C for 5 or 15 min
4 °C hold

The final incubation at 72 °C favors nontemplated nucleotide addition (Clark 1988, Kimpton et al., 1993). This final incubation temperature was dropped to 60 °C for some experiments in an effort to drive the nontemplated addition even further. Later experiments, including all of the larger sample sets, involved using the following thermal cycling program with TaqGold DNA polymerase:

95 °C for 11 min (to activate the TaqGold DNA polymerase)
40 cycles:
    94 °C for 30 sec
    55 °C for 30 sec
    72 °C for 30 sec
60 °C for 15 min
4 °C hold

Primers were typically designed to have an approximate annealing temperature of 57–63 °C and thus worked well with a 55 °C anneal step under this "universal" thermal cycling protocol. The need for extensive optimization of primer sets, reaction components, or cycling parameters was greatly reduced or eliminated with this approach for primer development on STR markers. Mitochondrial DNA samples were amplified with 35 cycles and an annealing temperature of 60 °C using the PCR primers listed in exhibit 26.

Multiplex PCR

Multiplex PCR was performed using a universal primer tagging approach (Shuber et al., 1995; Ross et al., 1998b) and the following cycling program:

95 °C for 10 min
50 cycles:
    94 °C for 30 sec
    55 °C for 30 sec
    68 °C for 60 sec
72 °C for 5 min
4 °C hold

The PCR master mix contained 5 mM MgCl$_2$, 2 U AmpliTaq Gold with 1X

Exhibit 26. **Mitochondrial DNA primers used for 10-plex SNP reaction**

Primer Name	SNP Site Position	SNP Base*	Sequence (5' → 3')	Cleaved Mass (Da)
MT5	H16224	A/G	b-GGAGTTGCAGTTGATGTGTGA(T)AGTTG	1,580
MT8'	L00146	T/C	b-GTCGCAGTATCTGTCTTTGAT(T)CCTGCC	1,790
MT10	H00247	C/T	b-CTGTGTGGAAAGTGGCTC(T)GCAGACATT	2,785
MT3'	H16189	A/G	b-GGTTGATTGCTGTACTTGCTTG(T)AAGCATGGGG	3,179
MT9	H00152	A/G	b-CTGTAATATTGAACGTAGG(T)GCGATAAATAAT	3,740
MT6	H16311	A/G	b-GTGCTATGTACGG(T)AAATGGCTTTATGT	4,355
MT2g	H16129	C/T	b-GTACTACAGGTGG(T)CAAGTATTTATGGTAC	4,957
MT1	H16069	G/A	b-AAATACA(T)AGCGGTTGTTGATGGGT	5,375
MT7	H00073	T/C	b-CCAGCGTC(T)GCGAATGCTATCGCGTGCA	5,891
MT4e	L00195	T/C	b-CTACGT(T)CAATATTACAGGCGAACATAC	6,500
DLOOP-F1**	L15997		CACCATTAGCACCCAAAGCT	
DLOOP-R1**	H00401		CTGTTAAAAGTGCATACCGCCA	

Note: "b" is listed for biotin; parentheses ( ) indicate the cleavable base.
*Anderson reference sequence base listed first.
**PCR Primers produced a 1,021 bp PCR product spanning positions 15977–00422.

Exhibit 27. **Multiplex PCR primers used for Y SNP markers.** Universal sequences have been attached to the 5'-end of the primers.

Primer Name	Primer Sequence (5' → 3')
M2-F3u	ATT TAG GTG ACA CTA TAG AAT ACG ACC CAG GAA GGT CCA GTA A
M2-R3u	TAA TAC GAC TCA CTA TAG GGA GAC CCC CTT TAT CCT CCA CAG AT
M3-F1u	ATT TAG GTG ACA CTA TAG AAT ACC TGC CAG GGC TTT CAA ATA G
M3-R1u	TAA TAC GAC TCA CTA TAG GGA GAC TGA AAT TTA AGG GCA TCT TTC A
M13-F1u	ATT TAG GTG ACA CTA TAG AAT ACT TAT GCC CAG GAA TGA ACA AG
M13-R1u	TAA TAC GAC TCA CTA TAG GGA GAC CCA TGA TTT TAT CCA ACC ACA TT
M119-F1u	ATT TAG GTG ACA CTA TAG AAT ACG GAA GTC ACG AAG TGC AAG T
M119-R1u	TAA TAC GAC TCA CTA TAG GGA GAC GGG TTA TTC CAA TTC AGC ATA CAG
M35-F1u	ATT TAG GTG ACA CTA TAG AAT ACA GGG CAT GGT CCC TTT CTA T
M35-R5u	TAA TAC GAC TCA CTA TAG GGA GAC TGG GTT CAA GTT TCC CTG TC
M55-F1u	ATT TAG GTG ACA CTA TAG AAT ACC AAA TAG GTG GGG CAA GAG A
M55-R1u	TAA TAC GAC TCA CTA TAG GGA GAC CCT GGG ATT GCA TTT GTA CTT
M60-F1u	ATT TAG GTG ACA CTA TAG AAT ACC CAA CAC TGA GCC CTG ATG
M60-R1u	TAA TAC GAC TCA CTA TAG GGA GAC GAG AAG GTG GGT GGT CAA GA
M42-F1u	ATT TAG GTG ACA CTA TAG AAT ACA GAT CAC CCA GAG ACA CAC AAA
M42-R1u	TAA TAC GAC TCA CTA TAG GGA GAC GCA AGT TAA GTC ACC AGC TCT C
M67-F1u	ATT TAG GTG ACA CTA TAG AAT ACG ACA AAC TCC CCT GCA CAC T
M67-R1u	TAA TAC GAC TCA CTA TAG GGA GAC CTT GTT CGT GGA CCC CTC TA
M69-F1u	ATT TAG GTG ACA CTA TAG AAT ACA CTC CTG GGT AGC CTG TTC A
M69-R1u	TAA TAC GAC TCA CTA TAG GGA GAC GAA CCA GAG GCA AGG GAC TA
M26-F1u	ATT TAG GTG ACA CTA TAG AAT ACC ACA GCA GAA GAG ACC AAG ACA
M26-R1u	TAA TAC GAC TCA CTA TAG GGA GAC TGG GGC TGT ATT TGA CAT GA
M96-F1u	ATT TAG GTG ACA CTA TAG AAT ACT GCC CTC TCA CAG AGC ACT T
M96-R1u	TAA TAC GAC TCA CTA TAG GGA GAC AGA TTC ACC CAC CCA CTT TG
M122-F2u	ATT TAG GTG ACA CTA TAG AAT ACA GTT GCC TTT TGG AAA TGA AT
M122-R2u	TAA TAC GAC TCA CTA TAG GGA GAC GGT ATT CAG CGC ATG CTG AT
M145-F1u	ATT TAG GTG ACA CTA TAG AAT ACG CTG GAG TCT GCA CAT TGA T
M145-R1u	TAA TAC GAC TCA CTA TAG GGA GAC TGG ATC ATG GTT CTT GAT TAG G
M45-F2u	ATT TAG GTG ACA CTA TAG AAT ACC ATC GGG GTG TGG ACT TTA C
M45-R2u	TAA TAC GAC TCA CTA TAG GGA GAC ACA GTG GCA CCA AAG GTC AT
M9-F1u	ATT TAG GTG ACA CTA TAG AAT ACA CTG CAA AGA AAC GGC CTA A
M9-R1u	TAA TAC GAC TCA CTA TAG GGA GAC TTT TGA AGC TCG TGA AAC AGA
M89-F4u	ATT TAG GTG ACA CTA TAG AAT ACC CAA ACA GCA AGG ATG ACA A
M89-R4u	TAA TAC GAC TCA CTA TAG GGA GAC TGC AAC TCA GGC AAA GTG AG
M17-F1u	ATT TAG GTG ACA CTA TAG AAT ACC TGG TCA TAA CAC TGG AAA TCA G
M17-R1u	TAA TAC GAC TCA CTA TAG GGA GAC CCA CTT AAC AAA CCC CAA AAT
M130-F1u	ATT TAG GTG ACA CTA TAG AAT ACG GGC AAT AAA CCT TGG ATT TC
M130-R1u	TAA TAC GAC TCA CTA TAG GGA GAC GCA ATT TAG CCA CTG CTC TG

Univ-F: ATT TAG GTG ACA CTA TAG AAT AC
Univ-R: TAA TAC GAC TCA CTA TAG GGA GAC

Exhibit 28. **CE electropherograms of "dropout" experiments conducted on a 9-plex PCR primer set used in developing Y-chromosome SNP markers.** Section (A) demonstrates that by simply removing 4 primer pairs (M130, M35, M122, and M145), the multiplex PCR yield improves (i.e., there are fewer remaining primers and the amplicon yields are more balanced). Panel (B) shows that by removing only M130 and M35, the remaining primers are reduced to the greatest extent (see red oval).

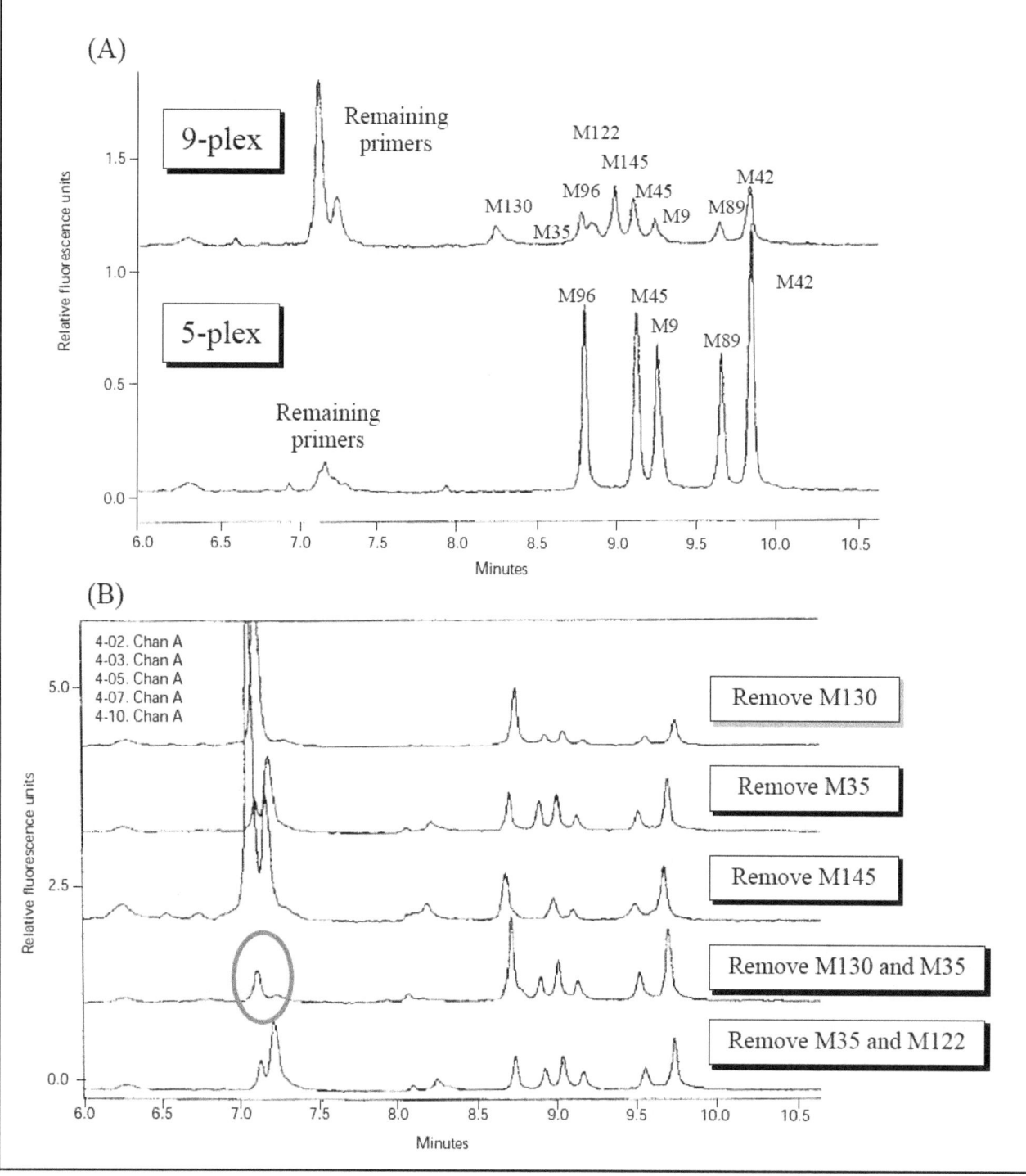

PCR buffer II (ABI), 20 pmol of each universal primer, and 0.2 pmol of each locus-specific primer. The universal primer sequences were 5'-ATTTAGGT-GACACTATAGAATAC-3' (attached on 5' end of locus specific forward primers) and 5'-TAATACGACTCAC-TATAGGGAGAC-3' (attached on 5' end of locus specific reverse primers). Exhibit 27 shows the primer sequences used for multiplex amplification of up to 18 Y SNP markers. During multiplex PCR development studies, each primer set was tested individually as well as in the multiplex set. Primer sets that were less efficient exhibited a higher amount of remaining primers or primer dimers in CE electrophero-grams of the PCR products. "Drop-out" experiments, where one or more primers were removed from the multiplex set, were then conducted to see which primer sets interfered with one another (exhibit 28). Finally, primer concentrations were adjusted to try and improve the multiplex PCR product balance between amplicons.

Verification of PCR amplification

Following PCR, a 1 μL aliquot of the PCR product was typically checked on a 2% agarose gel stained with ethidium bromide to verify amplification success. After a set of primers had been tested multiple times and a level of confidence had been gained for amplifying a particular STR locus, the gel PCR confirmation step was no longer used.

Later in this project, a Beckman P/ACE 5500 capillary electrophoresis (CE) instrument was used to check samples after PCR. The quantitative capabilities of CE are especially important when optimizing a multiplex PCR reaction. As long as the products are resolvable, their relative peak area or heights can be used to estimate amplification efficiency and balance during the multiplex PCR reaction. The CE separations were all performed using an intercalating dye

and sieving polymer solution as previously described (Butler et al., 1995) to avoid having to fluorescently label the PCR products. Samples were prepared for CE analysis by simply diluting a 1 μL aliquot of the amplicon in 49 μL of deionized water.

SNP reaction and phosphatase treatment

For SNP samples, the amplicons were treated with shrimp-alkaline phosphatase (SAP) (Amersham Pharmacia Biotech, Inc., Piscataway, NJ) to hydrolyze the unincorporated dNTPs following PCR (Haff and Smirnov, 1997). Typically, 1 U of SAP was added to each 20 μL PCR reaction and then incubated at 37 °C for 60 minutes followed by heating at 75 °C for 15 minutes. The SNP extension reaction consisted of a 5 μL aliquot of the SAP-treated PCR product, 1X TaqFS buffer, 1.2–2.4 U TaqFS (ABI), 12.5 μM dideoxynucleotide triphosphate (ddNTP) mix, and 0.5 μM biotinylated, cleavable SNP primer in a 20 μL volume. For multiplex analysis, SNP primer concentrations were balanced empirically, typically in the range of 0.3–1.5 μM, and polymerase and ddNTP concentrations were also doubled from the singleplex conditions to facilitate extension from multiple primers. The SNP extension reaction was performed in a thermal cycler with the following conditions: 94 °C for 1 min and 25–35 cycles at 94 °C for 10 sec, 45–60 °C (depending on the annealing temperature of the SNP primer) for 10 sec, and 70 °C for 10 sec. An annealing temperature of 52 °C was used for the mtDNA 10plex SNP assay.

Sample Cleanup and Mass Spectrometry

Following PCR amplification, a purification procedure involving solid-phase capture and release from streptavidin-coated magnetic beads was utilized (Monforte et al., 1997) to remove

salts that interfere with the MALDI ionization process (Shaler et al., 1996). At the start of this project, most of the sample purification was performed manually in 0.6 mL tubes with a 1.5 mL Dynal MPC®-E (Magnetic Particle Concentrator for Microtubes of Eppendorf Type) (Dynal A.S., Oslo, Norway). Larger scale experiments performed toward the end of this project utilized a robotic workstation fitted with a 96-tip pipettor that mimicked the manual method. This sample cleanup method involved washing the DNA with a series of chemical solutions to remove or reduce the high levels of sodium, potassium, and magnesium present from the PCR reaction. The PCR products were then released from the bead with a chemical cleavage step that breaks the covalent bond between the 5'-biotinylated portion of the DNA product and the remainder of the extension product, which contains the STR repeat region or the dideoxynucleotide added during the SNP reaction. In the final step prior to mass spectrometry analysis, samples were evaporated to dryness using a speed vac, reconstituted in 0.5 μL of matrix (manual protocol) or 2 μL of matrix (robotic protocol), and spotted on the sample plate.

The matrix typically used for STR analysis was a 5:1 molar ratio of 3-hydroxypicolinic acid (3-HPA) (Lancaster Synthesis, Inc., Windham, NH) with picolinic acid in 25 mM ammonium citrate (Sigma-Aldrich, St. Louis, MO) and 25% acetonitrile. For SNP analysis, ~0.5 M saturated 3-HPA was used with the same solvent of 25 mM ammonium citrate and 25% acetonitrile. A GeneTrace-designed and built linear time-of-flight mass spectrometer was used as previously described (Wu et al., 1994). Much of the early data were collected manually on a research mass spectrometer. During the time period of this project, GeneTrace also built multiple high-throughput instruments.

Automated high-throughput mass spectrometer

GeneTrace has designed and custom-built unique, automated time-of-flight mass spectrometers for high-throughput DNA analysis. The basic instrument design is covered under U.S. Patent 5,864,137 (Becker and Young, 1999). A high repetition rate UV laser (e.g., 100 Hz) is used to enable collection of high quality mass spectra consisting of 100–200 summed shots in only a few seconds. The sample chamber can hold up to two sample plates at a time with each plate containing 384 spotted samples. Exhibit 29 shows a sample plate on the X–Y table under the custom GeneTrace ion optics.

An important feature of this automated mass spectrometer is "peak picking" software that enables the user to define "good" versus "bad" mass spectra. After each laser pulse, the "peak picker" algorithm checks for peaks above a user-defined signal-to-noise threshold in a user-defined mass range. Only "good" spectra are kept and summed into the final sample spectrum, which improves the overall signal quality. The X–Y table moves in a circular pattern around each sample spot until either the maximum number of good shots (e.g., 200) or the maximum number of total shots (e.g., 1,000) is reached. A mass spectrum's signal-to-noise level is related to the number of laser shots collected. In general, signal-to-noise improves as the square root of the number of shots. Thus, improving the signal by a factor of two would require increasing the number of good shots collected by a factor of four.

Raw data files (.dat) were converted to "smoothed" data files (.sat) using custom software developed at Gene-Trace that improved data quality and involved several multipoint Savitzky-Golay averages along with a baseline subtraction algorithm (Carroll and Beavis, 1996). A set of samples was

Exhibit 29. **Photo of the automated GeneTrace mass spectrometry ion optics over a sample plate containing 384 different DNA samples.** A sample in the center of the plate is illuminated by a pulsing laser light.

collected under a single "header" file with identical peak picking parameters. Each header file recorded the mass calibration constants and peak picking parameters and listed all of the samples analyzed with the number of good shots collected versus the number of total shots taken for each sample.

Data points in mass spectrometry are collected in spectral channels that must be converted from a time value to a mass value. This mass calibration is normally performed with two oligonucleotides that span the mass range being examined. For example, a 36-mer (10,998 Da) and a 55-mer

(16,911 Da) were typically used when examining STRs in the size range of 10,000–40,000 Da. On the other hand, a 15-mer (4,507 Da) and its doubly charged ion (2,253.5 Da) were used to cover SNPs in the size range of 1,500–7,500 Da. Ideally, larger mass oligonucleotides would be used for STR analysis to obtain more accurate masses, but producing a clean, well-resolved peak above 25–30 kDa is a synthetic and instrumental challenge. The calibration was typically performed only once per day because the calibration remains consistent over hundreds of samples. The mass accuracy and precision are such that no sizing standards or allelic ladders need to be run to determine a sample's size or genotype (Butler et al., 1998b).

Delayed extraction (Vestal et al., 1995) and mass gating ("blanking") were used to improve peak resolution and sensitivity, respectively. Typically, a delay of 500–1,000 nanoseconds was used to eliminate ions below ~8,000 Da for STRs, and a delay of 250–500 nanoseconds was used with a signal blanking below ~1,000 Da for SNPs.

Sample Genotyping

Automated STR genotyping program (CallSSR or CallSTR)

During the time period of this project, GeneTrace developed an automated sample genotyping program, named CallSSR. The data sets described in the Results section were processed either with CallSSR version 1.82 or a modified version of the program named CallSTR. The program was written in C++ at GeneTrace by a scientific programmer named Nathan Hunt and can run on a Windows® NT platform. A reference DNA sequence is used to establish the possible STR alleles and their expected masses based on an expected repeat mass and range of alleles. This mass information is recorded in a mass ladder file (exhibit 30). In the case of the forensic STR loci examined in this project, the GenBank sequences were used as the reference DNA sequences.

Exhibit 30. **Mass ladder file for STR loci analyzed in this study.** Different masses exist for a locus due to different primer positions. The most commonly used primer sets are highlighted. The reference mass does not include nontemplate addition by the polymerase. The genotyping program automatically adds 313 Da to the reference mass for compensation of adenylation.

Locus Name	Repeat Mass (Da)	Reference Allele	Reference Mass (Da)	Minimum Repeat	Maximum Repeat	Forward Primer	Reverse Primer
TH01	1,260	9	21,133	3	12	TH01-F2	TH01-R2-GTS
TH01b	1,260	9	18,685	3	12	TH01-F	TH01-R-GTS
TH01b2	1,260	9	18,372	3	12	TH01-F (w/o+A)	TH01-R-GTS
TH01c	1,260	9	19,856	3	12	ddC	TH01-R2-GTS
TPOX	1,260	11	21,351	5	14	TPOX-F-GTS	TPOX-R
TPOXc	1,260	11	15,119	5	14	TPOX-F-GTS	ddC
CSF	1,260	12	28,890	5	16	CSF-F3-GTS	CSF-R3
CSFb	1,211	12	27,989	5	16	CSF-F3	CSF-R3-GTS
CSFc	1,260	12	22,993	5	16	CSF-F3-GTS	ddC
AMEL	951	6	27,333	3	7	AMEL-F-GTS	AMEL-R
AMELb	924	6	18,239	4	7	AMEL-F2	AMEL-R2-GTS
D7S820	1,260	12	21,639	5	16	D7S820-F-GTS	D7-R
D7S820c	1,260	12	16,949	5	16	D7S820-F-GTS	ddC
D3S1358	1,211	16	34,300	9	21	D3-F	D3-R-GTS
D3S1358b	1,260	15	25,624	9	21	D3-F2-GTS	D3-R2
D3S1358c	1,260	15	15,994	9	21	D3-F2-GTS	ddC
D16S539	1,211	11	18,180	4	16	D16-F	D16-R-GTS
D16S539b	1,260	11	26,668	4	16	D16-F4-GTS	D16-R4
D16S539c	1,260	11	14,765	4	16	D16-F4-GTS	ddC
D16S539d	1,211	11	25,572	4	16	D16-F4	D16-R4-GTS
D16S539e	1,211	11	19,127	4	16	D16-F	D16-R2-GTS
FGA	1,202	21	37,180	15	30	FGA-F2-GTS	FGA-R2
D8S1179	1,211	12	24,417	8	18	D8S1179-F-GTS	D8S1179-R
DYS391	1,211	9	23,004	7	14	DYS391-F-GTS	DYS391-R

CallSSR accepts as input smoothed, baseline-subtracted data files, a "layout" file, and the mass ladder file. The layout file describes each sample's position on the 384-well plate, the primer set used for PCR (i.e., the STR locus), and the DNA template name. The program processes samples at a rate of more than one sample per second so that a plate of 384 samples can be genotyped in less than 5 minutes. This high rate of processing speed is necessary in a high-throughput environment where thousands of samples must be genotyped every day.

Two files result from running the program: a "call file" and a "plot file." The call file may be imported into Microsoft® Excel for data examination and contains information like the allele mass and calculated sample genotype. The plot file generates plotting parameters that work with MATLAB (The Math Works, Inc., Natick, MA) scripts to plot 8 mass spectra per page, as seen in exhibit 31. Plots are generated in an artificial repeat space to aid visual inspection of the mass spectrometry data compared with allele bins. The CallSSR algorithm has been written to ignore stutter peaks and double-charged peaks, which are artifacts of the DNA amplification step and mass spectrometry ionization process, respectively.

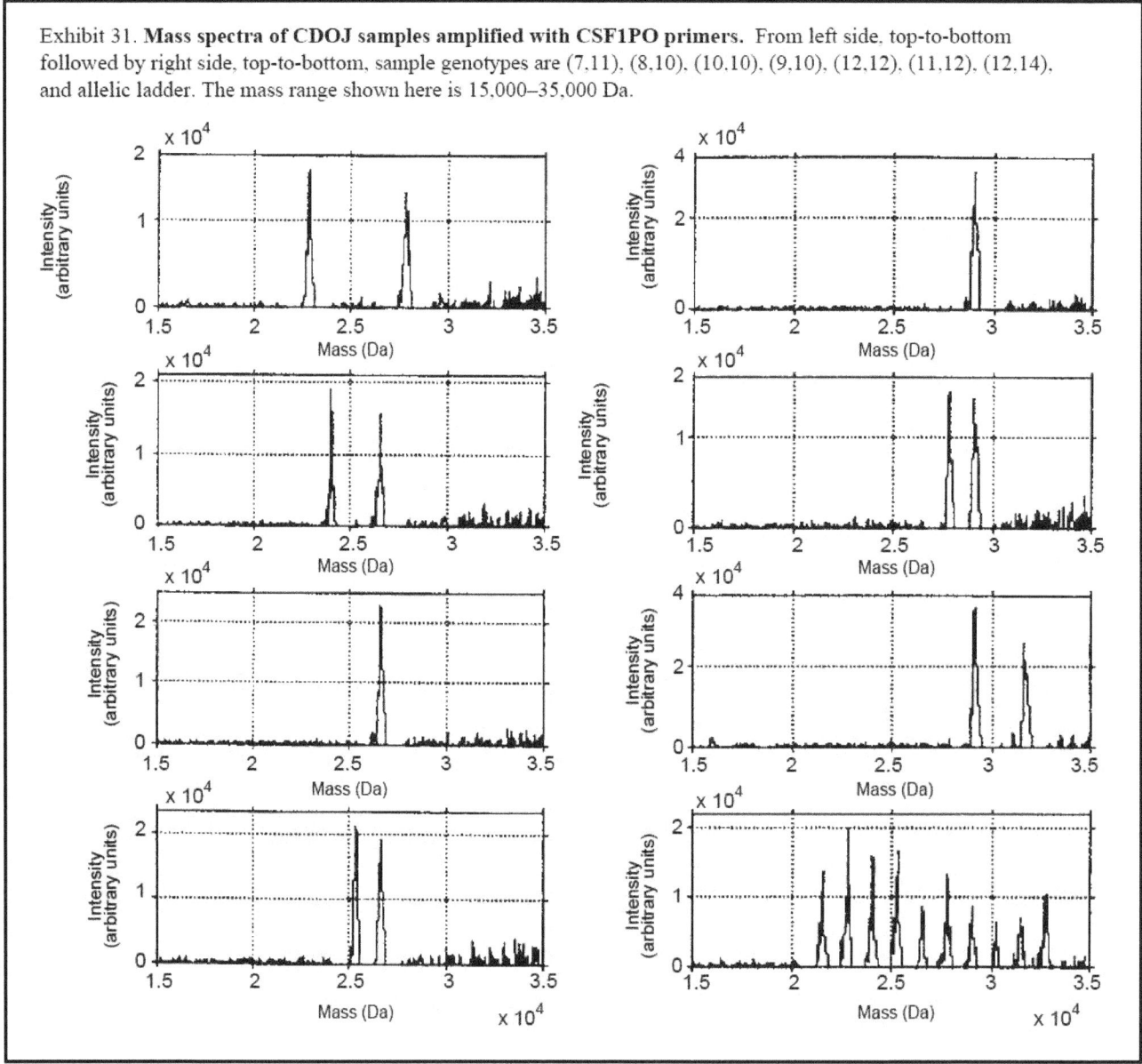

Exhibit 31. **Mass spectra of CDOJ samples amplified with CSF1PO primers.** From left side, top-to-bottom followed by right side, top-to-bottom, sample genotypes are (7,11), (8,10), (10,10), (9,10), (12,12), (11,12), (12,14), and allelic ladder. The mass range shown here is 15,000–35,000 Da.

Automated SNP genotyping program (CallSNP)

In-house automated SNP analysis software was developed and used to determine the genotype for each SNP marker. This program, dubbed CallSNP, was written in C++ by a scientific programmer named Kevin Coopman and will run on a Windows NT or UNIX platform. The software searches for an expected primer mass and, after locating the pertinent primer, searches for the four possible extension products by using a linear least squares fit, with the primer peak shape serving as the fitting line. In this way, peak adducts from the ionization process are distinguished from true heterozygotes. The fit coefficients of the four possible nucleotides are then compared with one another to determine the appropriate SNP base.

The base with the highest value (i.e., best fit) is the called base. The mass between the primer and the extension product can then be correlated to the incorporated nucleotide at the SNP site. In the case of a heterozygote at the SNP site, two extension products exist and are called by the software.

As with the CallSSR software, a layout file, a mass file, and mass spectrometry data files are required as input. Call files generate information regarding the closeness of the fit for each possible nucleotide with an error value associated for each call. The SNP mass information file includes the SNP marker name, expected primer mass (post-cleavage), and expected SNP bases. The current version of CallSNP works well for singleplex SNPs but needs modifications before it can work effectively on multiplex SNP samples. In

principle, the program could be scaled to limited, widely spaced multiplexes where the doubly charged ions of larger mass peaks do not fall in the range of lower mass primer peaks.

Comparison Tests With ABI 310 Genetic Analyzer

For comparison purposes, more than 200 genomic DNA samples were genotyped using the Applied Biosystems 310 Genetic Analyzer and the AmpF1STR® Profiler Plus™ or AmpF1STR® COfiler™ fluorescent STR kit. Exhibit 32 lists the numbers of samples analyzed with each STR kit. STR samples were run in the ABI 310 CE system using the POP-4 polymer, 1X Genetic Analysis buffer, and

Exhibit 32. **Sample sets run on ABI 310 Genetic Analyzer with AmpF1STR® Profiler Plus™ or AmpF1STR® COfiler™ fluorescent STR kits**

Sample Set	Number of Samples	Kit Used	MS Data Compared	Comments
DOJ plate	88	COfiler	TH01, TPOX, CSF1PO, D7S820, D3S1358, D16S539, D8S1179, FGA, DYS391, and amelogenin	*See exhibits 12–19, 39, and 40*
CEPH/Diversity plate	92	COfiler and ProfilerPlus	TH01, TPOX, CSF1PO, D7S820, D3S1358, D16S539, D8S1179, and amelogenin	*See exhibit 63*
Stanford male samples	37	ProfilerPlus	For Y SNP testing (not completed)	*See exhibits 44–45;* new microvariants in D18, D21, FGA seen
Butler family samples	34	COfiler	No mass spectrometry data collected	
JMB family samples	4	COfiler and ProfilerPlus	No mass spectrometry data collected	
Standard templates	3	COfiler and ProfilerPlus	TH01, TPOX, CSF1PO, D7S820, D3S1358, D16S539, D8S1179, FGA, DYS391, and amelogenin	*See exhibit 73* K562, AM209, and UP006
TOTAL TESTED	**258**	221 COfiler 136 ProfilerPlus	**2,907 total genotypes measured**	
CDOJ results	88	Profiler (under CDOJ validated method)	*See above*	B11, D3S1358:15,15.2 incorrectly called 14,15 *(see exhibit 74)*

a 47-cm (50 μm i.d.) capillary with the GS STR POP4 (1 mL) F separation module (ABI Prism 310 Genetic Analyzer User's Manual, 1998). With this module, samples were electrokinetically injected for 5 seconds at 15,000 volts and separated at 15,000 volts for 24 minutes with a run temperature of 60 °C. DNA sizing was performed with ROX-labeled GS500 as the internal sizing standard. Samples were prepared by adding 1 μL PCR product to 20 μL deionized formamide containing the ROX-GS500 standard. The samples were heat-denatured at 95 °C for 3 minutes and then snap cooled on ice prior to being loaded into the autosampler tray. These separation conditions and sizing standards are commonly used in validated protocols by forensic DNA laboratories. Following data collection, samples were analyzed with Genescan 2.1 and Genotyper 2.0 software programs (ABI).

While standard CE conditions were used, new PCR conditions were developed to dramatically reduce the cost of using ABI's STR kits. The PCR volume was reduced from the standard 50 μL described in the ABI protocol (AmpF1STR® ProfilerPlus™, 1998, and AmpF1STR® Cofiler™, 1998) to 5 μL, which corresponded with a cost reduction of 90% per DNA amplification. The kit reagents were mixed in their ABI-specified proportions—11 μL primer mix, 1 μL TaqGold polymerase (5 U/μL), and 21 μL PCR mix. A 3 μL aliquot of this master mix was then added to each tube along with 2 μL of genomic DNA template (typically at 1–2 ng/μL). Both PE9700 and MJ Research thermal cyclers worked for this reduced PCR volume method provided that the 200 μL PCR tubes were sealed well to prevent evaporation. Results showed that an 8-strip of 0.2 mL thin wall PCR tubes from Out Patient Services, Inc., (OPS) (Petaluma, CA) worked best for the PE9700 thermal cycler. Only 1 μL is needed for CE sample preparation; a sample that can be re-injected multiple times if needed. Thus, with a 50 μL PCR reaction, 49 μL were never used to produce a result under the standard ABI protocol. A 5 μL PCR produces less waste in addition to being less expensive. More importantly, the multiplex STR amplicons were more concentrated in a lower volume and produced higher signals in the ABI 310 data collection. Peak signals were often off-scale, and the number of cycles in the cycling program could be reduced from 28 to 26, or even 25 with some DNA templates. This 5 μL PCR also worked well with FTA paper punches that have been washed with FTA purification reagent (Life Technologies).

The California Department of Justice ran one plate of 88 samples with the AmpF1STR® Profiler™ kit on an ABI 310 Genetic Analyzer and provided those genotypes for comparison purposes. These results provided an independent verification of our work.

RESULTS AND DISCUSSION OF STR ANALYSIS BY MASS SPECTROMETRY

In the course of this work, thousands of data points were collected using STR markers of forensic interest verifying that GeneTrace's mass spectrometry technology works. During this same time, tens of thousands of data points were gathered across hundreds of different microsatellite markers from corn and soybean as part of an ongoing plant genomics partnership with Monsanto Company (St. Louis, MO). Whether the DNA markers used come from humans or plants, the characteristics described below apply when analyzing polymorphic repeat loci.

Marker Selection and Feasibility Studies With STR Loci

Prior to receiving grant funding, feasibility work had been completed using the STR markers TH01, CSF1PO, FES/FPS, and F13A1 in the summer and fall of 1996 (Becker et al., 1997). At the start of this project, a number of STR loci were considered as possible candidates to expand upon the initial four STR markers and to develop a set of markers that would work well in the mass spectrometer and would be acceptable to the forensic DNA community. Searches were made of publicly available databases, including the Cooperative Human Linkage Center (*http://lpg.nci.nih.gov/CHLC*), the Genome Database (*http://gdbwww.gdb.org*), and Weber set 8 of the Marshfield Medical Research Foundation's Center for Medical Genetics (*http://www.marshmed.org/genetics*). Literature was also searched for possible tetranucleotide markers with PCR product sizes below 140 bp in size to avoid having to redesign the PCR primers to meet our limited size range needs (Hammond et al., 1994; Lindqvist et al., 1996). The desired characteristics also included high heterozygosity, moderate number of alleles (<7 or 8 to maintain a narrow mass range) with no known microvariants (to avoid the need for a high degree of resolution), and balanced allele frequencies (most commonly allele <40% and least common allele >5%). This type of marker screen was found to be rather inefficient because the original primer sets reported in public STR databases were designed for gel-based separations, which were optimal over a size range of 100–400 bp. In fact, most of the PCR product sizes were in the 200–300 bp range. From a set of several thousand publicly available STRs, only a set of eight candidate tetranucleotide STRs were initially identified; three of which were tested using the original reported primers (exhibit 33). The initial goal was to identify ~25 markers that spanned all 22 autosomal chromosomes as well as the X and Y sex chromosomes.

Researchers quickly realized that population data were not available on these "new" markers and would not be readily accepted without extensive testing and validation. Since one of the objectives was to produce STR marker sets that would be of value to the forensic DNA community, the next step taken was the examination of STR markers already in use. After selecting

Exhibit 33. **Tetranucleotide markers identified through literature and public database searches as possible candidates for early STR marker development.** The databases searched included the Cooperative Human Linkage Center, Marshfield Clinic Weber set 8, and the Genome Database. Primers were synthesized and tested for the markers in **bold**.

Marker Name	Heterozygosity	Size Range	Allele Frequencies	Number of Alleles	GenBank Sequence
D1S1612	0.83	94–134 bp	0.9–26%	10 alleles	G07863
D2S1391	0.79	109–137 bp	5–35%	8 alleles	G08168
D5S1457	0.74	97–127 bp	1–32%	8 alleles	G08431
GATA132B04 (D5S2843)	Not reported	98–114 bp	12–35%	5 alleles	G10407
D16S2622	Not reported	71–91 bp	3–54%	5 alleles	G07934
D16S764	0.70	96–116 bp	4–38%	5 alleles	G07928
D19S591	0.74	96–112 bp	6–36%	5 alleles	G09745
D22S445	0.65	110–130 bp	3–37%	6 alleles	G08096

STR markers used by the Promega Corporation, Applied Biosystems, and the Forensic Science Service (FSS), researchers redesigned primer pairs for each STR locus to produce smaller PCR product sizes. These STR markers included TPOX, D5S818, D7S820, D13S317, D16S539, LPL, F13B, HPRTB, D3S1358, VWA, FGA, CD4, D8S1179, D18S51, and D21S11. The primers for TH01 and CSF1PO were also redesigned to improve PCR efficiencies and to reduce the amplicon sizes. Primers for amelogenin, a commonly used sex-typing marker, were also tested (Sullivan et al., 1993). In addition, two Y-chromosome STRs, DYS19 and DYS391, were examined briefly. Exhibit 34 summarizes the STR primer sets that were developed and tested over the course of this project. However,

with the announcement of the 13 CODIS core loci in the fall of 1997, emphasis switched to CSF1PO, TPOX, TH01, D3S1358, VWA, FGA, D5S818, D7S820, D13S317, D16S539, D8S1179, D18S51, D21S11, and the sex-typing marker amelogenin.

The newly designed GeneTrace primers produced smaller PCR products than those commercially available from Applied Biosystems or Promega (exhibit 2), yet resulted in identical genotypes in almost all samples tested. For example, correct genotypes were obtained on the human cell line K562, a commonly used control for PCR amplification success. Exhibit 35 shows the K562 results for CSF1PO, TPOX, TH01, and amelogenin. These results were included as part of a

publication demonstrating that time-of-flight mass spectrometry could perform accurate genotyping of STRs without allelic ladders (Butler et al., 1998).

Caveats of STR analysis by mass spectrometry

While mass spectrometry worked well for a majority of the STR markers tested, a few limitations excluded some STRs from working effectively. Two important issues that impact mass spectrometry results are DNA size and sample salts. Mass spectrometry resolution and sensitivity are diminished when either the DNA size or the salt content of the sample is too large (Ross and Belgrader, 1997; and Taranenko et al., 1998). By designing the PCR primers to bind close to the repeat region, the STR allele sizes are reduced so that resolution and sensitivity of the PCR products are benefited. In addition, the GeneTrace-patented cleavage step reduces the measured DNA size even further. When possible, primers are designed to produce amplicons that are less than 120 bp, although work is sometimes undertaken with STR alleles that are as large as 140 bp in size. This limitation in size prevents reliable analysis of STR markers with samples containing a large number of repeats, such as most of the FGA, D21S11, and D18S51 alleles (exhibit 2).

To overcome the sample salt problem, researchers used a patented solid-phase purification procedure that reduced the concentration of magnesium, potassium, and sodium salts in the PCR products prior to being introduced to the mass spectrometer (Monforte et al., 1997). Without the reduction of the salts, resolution is diminished by the presence of adducts. Salt molecules bind to the DNA during the MALDI ionization process and give rise to peaks that have a mass of the DNA molecule plus the salt molecule. Adducts broaden peaks and thus reduce peak resolution. The sample

Exhibit 34. **STR markers examined at GeneTrace during the course of this project as sorted by their chromosomal position.** Primers were designed, synthesized, and tested for each of these markers. The most extensive testing was performed with the markers highlighted in green. Amelogenin, which is a gender identification marker rather than an STR, is listed twice because it occurs on both the X and Y chromosomes. The *italicized* STRs are those not commonly used by the forensic DNA community.

Human Chromosome	STR Marker	Human Chromosome	STR Marker
1	F13B	13	D13S317
2	TPOX	14	
3	D3S1358	15	FES/FPS
4	FGA	16	D16S539, *D16S2622*
5	CSF1PO, D5S818, *GATA132B04*	17	
6	F13A1	18	D18S51
7	D7S820	19	
8	LPL, D8S1179	20	
9		21	D21S11
10		22	*D22S445*
11	TH01	X	HPRTB, Amelogenin
12	VWA, CD4	Y	DYS19, DYS391, Amelogenin

Exhibit 35. **Mass spectra for CSF1PO, TPOX, TH01, and amelogenin using K562 DNA.** Genotypes agree with results reported by the manufacturer (Promega Corporation). The numbers above the peaks represent the allele calls based upon the observed mass. The allele imbalance on the heterozygous samples is because the K562 strain is known to contain an unusual number of chromosomes and some of them are represented more than twice per cell. The TH01 peak is split because it is not fully adenylated (Butler et al., 1998).

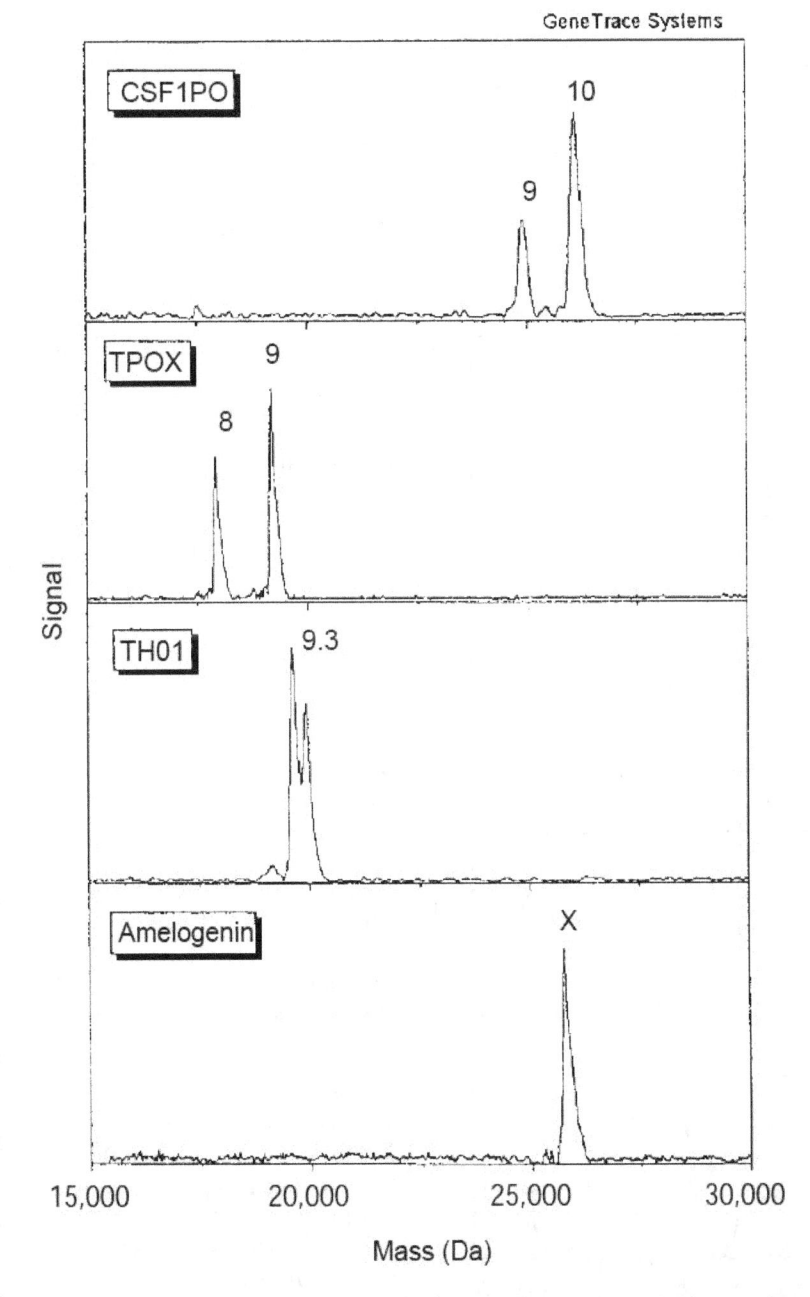

purification procedure, which was entirely automated on a 96-tip robotic workstation, reduced the PCR buffer salts and yielded "clean" DNA for the mass spectrometer. Appropriate care must be taken to prevent samples from being contaminated with salts both during and after the sample purification procedure.

Size reduction methods

The portion of the DNA product on the other side of the repeat region from the cleavable primer was removed in one of two possible ways: using a restriction enzyme (Monforte et al., 1999) or performing a nested linear amplification with a ddN terminating nucleotide (Braun et al., 1997a and 1997b). Both methods have pros and cons. A restriction enzyme, *Dpn*II, which recognizes the sequence 5'... ^GATC...3', was used with VWA samples to remove 45 bp from each PCR product. For example, the GenBank allele that contains 18 repeat units and is 154 bp following PCR amplification may be reduced to 126 nucleotides following primer cleavage, but it can be shortened to 81 nucleotides if primer cleavage is combined with *Dpn*II digestion. At 81 nt or 25,482 Da, the STR product size is much more manageable in the mass spectrometer. This approach works nicely provided the restriction enzyme recognition site remains unchanged. The *Dpn*II digestion of VWA amplicons worked on all samples tested, including a reamplification of an allelic ladder from ABI (exhibit 36). However, the cost and time of analysis are increased with the addition of a restriction enzyme step.

The second approach for reducing the overall size of the DNA molecule in the mass spectrometer involved using a single ddNTP with three regular dNTPs. A linear amplification extension reaction was performed with the ddNTP terminating the reaction on the opposite side of the repeat from the cleavable primer. However, there were several limitations with this "single base sequencing" approach. First, it only worked if the repeat did not contain all four nucleotides. For example, a nucleotide mixture of dideoxycytosine (ddC), deoxyadenosine (dA), deoxythymidine (dT), and deoxyguanosine (dG) will allow extension through an AATG repeat (as occurs in the bottom strand of TH01) but will terminate at the first C nucleotide in a TCAT repeat (the top strand of TH01). Thus one is limited with the DNA strand that can be used for a given combination of dideoxynucleotide and corresponding deoxynucleotides. In addition, primer position and STR sequence content are important. If a ddC mix is used, the DNA sample cannot contain any C nucleotides prior to the repeat region or within the repeat, or the extension will prematurely halt and the information content of the full repeat will not be accurately captured. In most cases, this requires the extension primer to be immediately adjacent to the STR repeat, a situation that is not universally available due to the flanking sequences around the repeat region. For example, this approach will work with TH01 (AATG) but not VWA, which has three different repeat structures: AGAT, AGAC, and AGGT. Thus with VWA, a ddC would extend through the AGAT repeat but would be prematurely terminated at the C in the AGAC repeat, and valuable polymorphic information would be lost.

The use of a terminating nucleotide also provides a sharper peak for an amplified allele compared with the split peaks or wider peaks (if resolution is poor) that can result from partially adenylated amplicons (i.e., -A/+A). Exhibit 37 illustrates the advantage of a ddG termination on a D8S1179 heterozygous sample containing 11 and 13 TATC repeats. In the bottom panel, 23 nt were removed compared with the top panel, which corresponds to a mass reduction of almost 8,000 Da. The peaks are sharper in the lower panel, as the products are blunt ended. Identical genotypes were obtained with both approaches, illustrating that the ddG termination is occurring at the same point on the two different sized alleles.

To summarize, STR sample sizes were reduced using primers that have been designed to bind close to the repeat region or even partially on the repeat itself. A cleavable primer was incorporated into the PCR product to allow post-PCR chemical cleavage and subsequent mass reduction. Two additional post-PCR methods were also explored to further reduce the measured DNA size. These methods

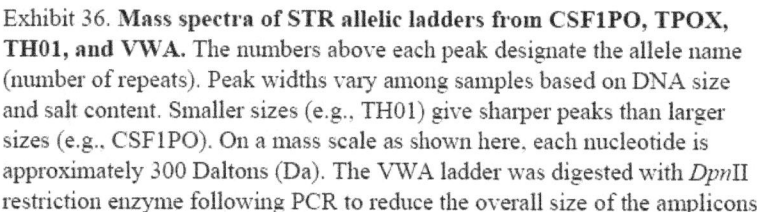

Exhibit 36. **Mass spectra of STR allelic ladders from CSF1PO, TPOX, TH01, and VWA.** The numbers above each peak designate the allele name (number of repeats). Peak widths vary among samples based on DNA size and salt content. Smaller sizes (e.g., TH01) give sharper peaks than larger sizes (e.g., CSF1PO). On a mass scale as shown here, each nucleotide is approximately 300 Daltons (Da). The VWA ladder was digested with *Dpn*II restriction enzyme following PCR to reduce the overall size of the amplicons.

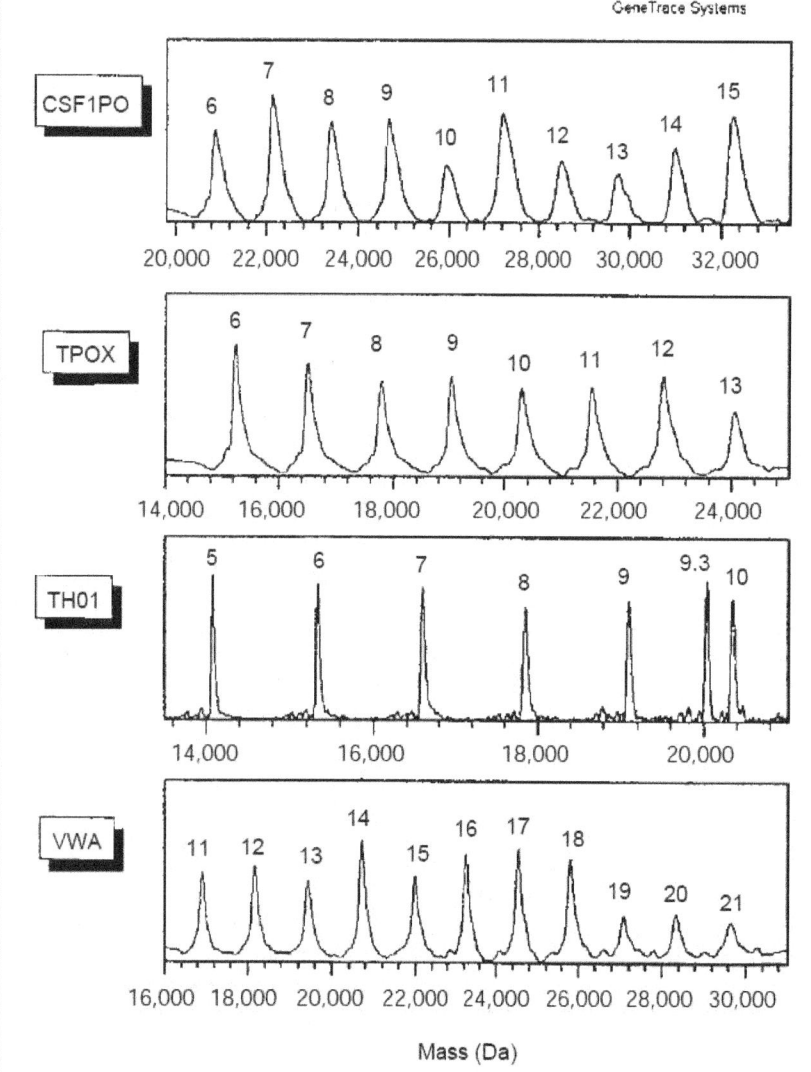

Exhibit 37. **Mass spectra of a D8S1179 sample illustrating the benefit of a dideoxynucleotide termination approach.**
The top panel displays a result from a regular PCR product; the bottom panel contains the same sample treated with a linear
amplification mix containing a ddG terminator with dA, dT, and dC deoxynucleotides. With the ddG approach, the problem
of incomplete adenylation (both –A and +A forms of a PCR product) is eliminated, and the amplicons are smaller, which
improves their sensitivity and resolution in the mass spectrometer. The red oval highlights the broader -A/+A peaks present
in the regular PCR product. Note: The genotype (i.e., 11 and 13 repeats) is identical between the two approaches, even
though almost 8,000 Da are removed with the ddG termination.

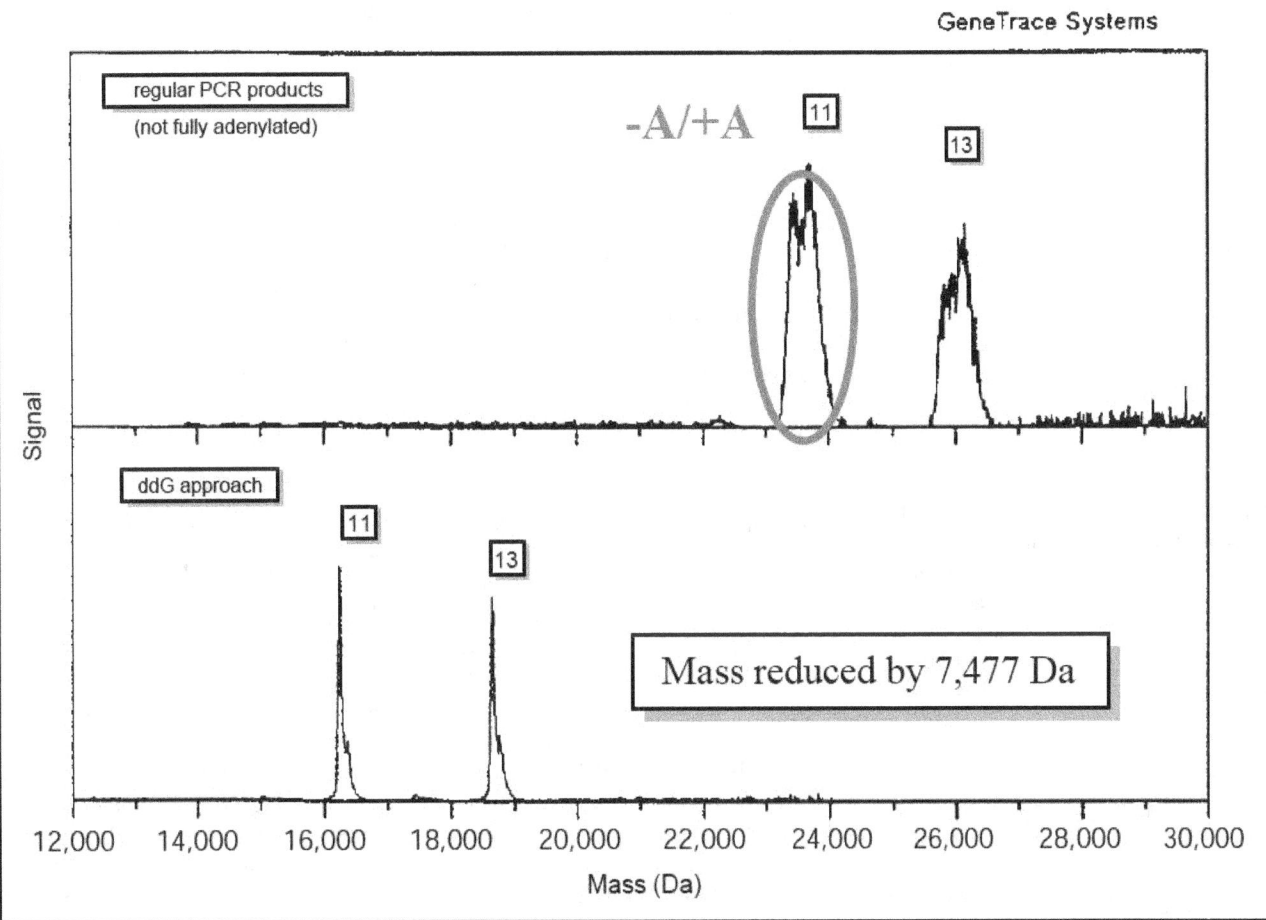

included restriction enzyme digestion in the flanking region on the other side of the repeat region from the cleavable primer and a primer extension through the repeat region with a single dideoxynucleotide terminator (single base sequencing approach).

To illustrate the advantages of these approaches to reduce the overall DNA product mass, researchers examined the STR locus TPOX. Using a conventional primer set, a sample containing 11 repeats measured 232 bp or ~66,000 Da. By redesigning the primers to anneal close to the repeat region, a PCR product of 89 bp was obtained. With the cleavable primer, the size was reduced to 69 nt or 21,351 Da. By incorporating a ddC termination reaction, another 20 nt were removed leaving only 49 nt or ~12,000 Da (primarily only the repeat region). The repeat region contained 44 nt (4 nt x 11 repeats) or ~10,500 Da. The ddC termination was also used in multiplex STR analysis to produce a CSF1PO-TPOX-TH01 triplex (exhibits 4 and 5). The repeat sequences used for these STR loci were AGAT for CSF1PO, AATG for TPOX, and AATG for TH01. The level of sequence clipping by ddC was as follows: CSF1PO (-14 nt), TPOX (-20 nt), and TH01 (-4 nt).

Multiplex STR Work

Due to the limited size range of DNA molecules that may be analyzed by this technique, a new approach to multiplexing was developed that involved interleaving alleles from different loci rather than producing nonoverlapping multiplexes. If the amplicons could be kept under ~25,000 Da, a high degree of mass accuracy and resolution could be used to distinguish alleles from multiple loci that may differ by only a fraction of a single nucleotide (exhibit 10). Allelic ladders are useful to demonstrate that all alleles in a multiplex are distinguishable (exhibit 9).

The expected masses for a triplex involving the STR loci CSF1PO, TPOX, and TH01 (commonly referred to as a CTT multiplex) are schematically displayed in exhibit 4. All known alleles for these STR loci, as defined by STRBase (Ruitberg et al., 2001), are fully resolvable and far enough apart to be accurately determined. For example, TH01 alleles 9.3 and 10 fall between CSF1PO alleles 10 and 11. For all three STR systems in this CTT multiplex, the AATG repeat strand is measured, which means that the alleles *within* the same STR system differ by 1,260 Da. The smallest spread between alleles *across* multiple STR systems in this particular multiplex exists between the TPOX and TH01 alleles, where the expected mass difference is 285 Da. TPOX and CSF1PO alleles differ by 314 Da, while TH01 and CSF1PO alleles differ by 599 Da. By using the same repeat strand in the multiplex, the allele masses between STR systems all stay the same distance apart. Each STR has a unique flanking region and it is these sequence differences between STR systems that permit multiplexing in such a fashion as described here. An actual result with this CTT multiplex is shown in exhibit 5. This particular sample is homozygous for both TPOX (8,8) and CSF1PO (12,12) and heterozygous at the TH01 locus (6,9.3).

It is also worth noting that this particular CTT multiplex was designed to account for possible, unexpected microvariants. For example, a CSF1PO allele 10.3 that appears to be a single base shorter than CSF1PO allele 11 was recently reported (Lazaruk et al., 1998). With the CTT multiplex primer set described here, a CSF1PO 10.3 allele would have an expected mass of 21,402 Da, which would be fully distinguishable from the nearest possible allele (i.e., TH01 allele 10) because these alleles would be 286 Da apart. Using a mass window of 100 Da as defined by previous precision studies (Butler et al., 1998), all possible alleles including microvariants should be fully distinguishable. STR multiplexes are designed so that expected allele masses between STR systems are offset in a manner that possible microvariants, which are most commonly insertions or deletions of a partial repeat unit, may be distinguished from all other possible alleles. The larger the allele's mass range, the more difficult it becomes to maintain a high degree of mass accuracy. For example, exhibit 8 shows the observed mass for TH01 allele 9.3 is -52 Da from its expected mass, while TPOX allele 9 is only 3 Da from its expected mass. In this particular case, the mass calibrants used were 4,507 Da and 10,998 Da. Thus, the TPOX allele's mass measurement was more accurate and closer to the calibration standard. The ability to design multiplexes that have a relatively compact mass range is important to maintaining the high level of mass accuracy needed for closely spaced alleles from different, overlapping STR loci. The mass calibration standards should also span the entire region of expected measurement to guarantee the highest degree of mass accuracy.

Two possible multiplexing strategies for STR genotyping are illustrated in exhibit 38. Starting with a single punch of blood stained FTA paper, it is possible to perform a multiplex PCR (simultaneously amplifying all STRs

of interest) followed by another PCR with primer sets that are closer to the repeat region. With this approach, single or multiplexed STR products can be produced that are small enough for mass spectrometry analysis. Alternatively, multiple punches could be made from a single bloodstain on the FTA paper followed by singleplex or multiplex PCR with mass spectrometry primers. After the genotype is determined for each STR locus in a sample, the information would be combined to form a single sample genotype for inclusion in CODIS or some other DNA database. This multiplexing approach permits flexibility for adding new STR loci or only processing a few STR markers across a large number of samples at a lower cost than processing extensive and inflexible STR multiplexes.

Comparison Tests Between ABI 310 and Mass Spectrometry Results

A plate of 88 samples from the CDOJ DNA Laboratory was tested with 10 different STR markers and compared with results obtained using the ABI 310 Genetic Analyzer and commercially available STR kits. The samples were supplied as a 200 µL aliquot of extracted genomic DNA in a 96-well tray with each sample at a concentration of 1 ng/µL. A 5 µL aliquot was used for each PCR reaction, or 5 ng total per reaction. Since each marker was amplified and examined individually, approximately 35 ng of extracted genomic DNA was required to obtain genotypes on the same 7 markers as were amplified in a single AmpF1STR® COfiler™ STR multiplex. Only 2 ng of genomic DNA were used per reaction with the AmpF1STR® COfiler™ kit. Thus, a multiplex PCR reaction is much better suited for situations where the quantity of DNA is limited (e.g., crime scene sample). However, in most cases involving high-throughput DNA typing (e.g., offender database work), hundreds of nanograms of extracted DNA would be easily available.

A major advantage of the mass spectrometry approach is speed of the technique and the high-throughput capabilities when combined with robotic sample preparation. The data collection times required for the 88 CDOJ samples using the ABI 310 Genetic Analyzer and GeneTrace's mass spectrometry method are compared in exhibit 11. While it took the ABI 310 almost 3 days to collect the data for the 88 samples, the same genotypes were obtained on the mass spectrometer in less than 2 hours. Even the ability to analyze multiple STR loci simultaneously with different fluorescent tags on the ABI 310 could not match the speed of GeneTrace's mass spectrometry data collection with each marker run individually.

To verify that the mass spectrometry approach produces accurate results, comparison studies were performed on the genotypes obtained from the two different methods across 8 different STR loci. Exhibits 12–19 contain a direct comparison with 1,408 possible data points (2 methods × 88 samples × 8 loci). With a few minor exceptions, there was almost a 100% correlation between the two methods. In addition to the data obtained on the 8 loci from both the ABI 310 and the mass spectrometer, two additional markers (D8S1179 and DYS391) were measured by mass spectrometry across these same 88 samples (exhibits 39–40). Both the D8S1179 and the DYS391 primer sets worked extremely well in the mass spectrometer (exhibit 41). Thus, it is likely that if results were made available on these same samples with fluorescent STR primer sets (e.g., D8S1179 is in the AmpF1STR® Profiler Plus™ kit), there would also be a further correlation between the two methods.

PCR Issues

Null alleles

When making comparisons between two methods that use different PCR primer sets, the issue is whether or not a different primer set for a given STR locus will result in different allele calls through possible sequence polymorphisms in the primer binding sites. In other words, do primers used for mass spectrometry that are closer to the repeat region than those primers used

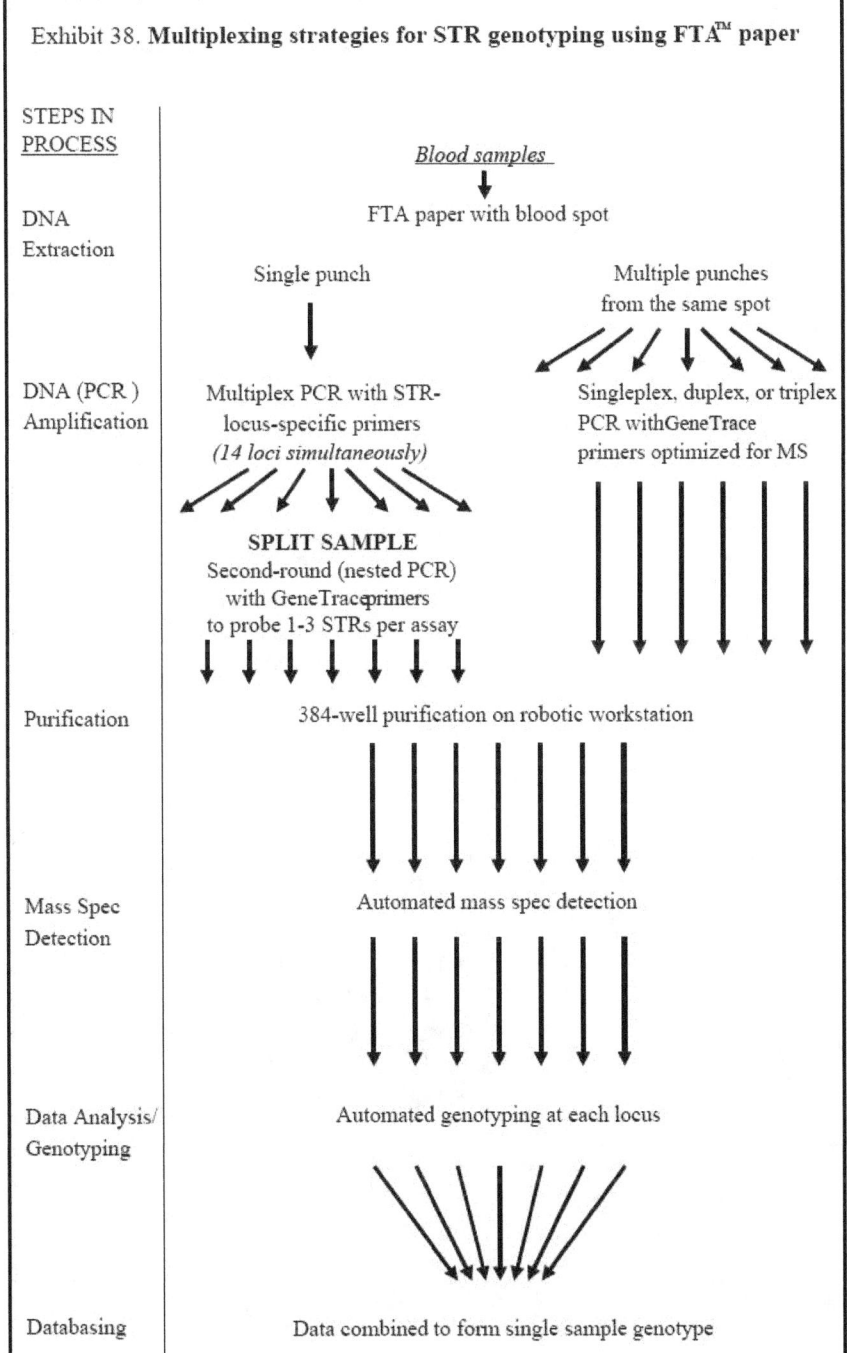

Exhibit 38. **Multiplexing strategies for STR genotyping using FTA™ paper**

STEPS IN PROCESS

DNA Extraction

Blood samples

FTA paper with blood spot

Single punch

Multiple punches from the same spot

DNA (PCR) Amplification

Multiplex PCR with STR-locus-specific primers _(14 loci simultaneously)_

Singleplex, duplex, or triplex PCR with GeneTrace primers optimized for MS

SPLIT SAMPLE
Second-round (nested PCR) with GeneTrace primers to probe 1-3 STRs per assay

Purification

384-well purification on robotic workstation

Mass Spec Detection

Automated mass spec detection

Data Analysis/ Genotyping

Automated genotyping at each locus

Databasing

Data combined to form single sample genotype

Exhibit 39. **CDOJ D8S1179 STR results with the mass spectrometry method**

Position	Mass Spec	Allele 1 (Da)	Allele 2 (Da)	Position	Mass Spec	Allele 1 (Da)	Allele 2 (Da)
A1	12,15	24,693	28,228	A7	13,13	25,791	
B1	14,15	27,159	28,369	B7	13,14	25,845	26,954
C1	11,14	23,342	26,933	C7	12,15	24,540	28,101
D1	15,15	28,199		D7	13,14	25,772	26,931
E1	11,14	23,421	26,988	E7	14,14	26,960	
F1	13,17	25,822	30,529	F7	13,16	25,785	29,381
G1	11,14	23,449	27,093	G7	14,14	27,097	
H1	14,14	27,067		H7	14,14	27,039	
A2	13,13	25,777		A8	13,15	25,770	28,204
B2	12,14	24,597	27,007	B8	11,13	23,383	25,816
C2	11,11	23,361		C8	14,15	27,033	28,043
D2	11,13	23,399	25,768	D8	14,14	27,076	
E2	13,14	25,904	27,012	E8	14,15	26,984	28,221
F2	14,14	27,125		F8	13,14	25,781	26,946
G2	No data			G8	10,12	22,183	24,581
H2	11,15	23,344	28,151	H8	12,15	24,587	28,197
A3	15,15	28,160		A9	15,16	28,265	29,390
B3	13,15	25,904	28,256	B9	13,14	25,835	26,963
C3	12,14	24,640	27,022	C9	11,14	23,447	27,084
D3	11,15	23,407	28,201	D9	11,14	23,391	26,988
E3	11,14	23,461	27,093	E9	11,14	23,361	26,931
F3	14,15	27,086	28,186	F9	14,15	27,031	28,225
G3	10,15	22,218	28,230	G9	14,14	26,978	
H3	13,16	25,849	29,355	H9	11,13	23,451	26,008
A4	14,15	27,018	28,112	A10	13,14	25,789	26,841
B4	12,14	24,605	27,018	B10	15,15	28,206	
C4	13,13	25,797		C10	13,13	25,791	
D4	No data			D10	14,15	26,999	28,123
E4	14,14	27,108		E10	13,14	25,895	26,980
F4	14,15	27,080	28,215	F10	13,15	25,777	28,197
G4	12,13	24,311	25,519	G10	14,15	26,733	27,909
H4	13,14	25,824	27,044	H10	12,14	24,688	27,093
A5	12,16	24,548	29,321	A11	12,15	24,574	28,230
B5	13,14	25,893	26,997	B11	13,14	25,822	27,054
C5	12,13	24,727	25,885	C11	14,16	26,963	29,324
D5	No data			D11	13,14	25,812	26,950
E5	14,14	26,960		E11	12,15	24,617	28,232
F5	13,14	25,916	27,014	F11	14,14	26,982	
G5	14,15	27,009	28,106	G11	14,14	27,005	
H5	13,14	25,804	26,708	H11	12,13	24,631	25,818
A6	14,14	26,988					
B6	14,16	26,982	29,364				
C6	15,15	28,249					
D6	14,14	26,716					
E6	13,16	25,933	29,584				
F6	8,14	19,884	27,041				
G6	15,15	28,394					
H6	14,14	27,009					

in fluorescent STR typing yield the same genotype?

Differences between primer sets are possible if there are sequence differences outside the repeat region that occur in the primer binding region of either set of primers (exhibit 42). This phenomenon produces what is known as a "null" allele, or in other words, the DNA template exists for a particular allele but fails to amplify during PCR due to primer hybridization problems. In all cases except the STR locus D7S820, there was excellent correlation in genotype calls between the two methods (where mass spectrometry and CE results were obtained), signifying that the mass spectrometry primers did not produce any null alleles.

For the STR locus D7S820, 17 of 88 samples did not agree with the two methods (exhibit 18). The bottom two panels in exhibit 43 illustrate more microheterogeneity at this locus than previously reported. On the lower left plot, only the allele 10 peak can be seen; allele 8, which was seen with PCR amplification using a fluorescent primer set, is missing (see position of red arrow in exhibit 43). On the lower right plot, both allele 8 and allele 10 are amplified and detected in the mass spectrometer, confirming that the problem is with the PCR amplification and not the mass spectrometry data collection. In this particular case, there is a difference between those two alleles 8, meaning that the mass spectrometer primer set identified a new, previously unreported allele. When using fluorescent primer sets that anneal 50–100 bases or more from the repeat region, a single-base change (e.g., T to C) out of a 300 bp PCR product is difficult to detect. Upon comparing the results of mass spectrometer data where there were missing alleles with the results from the ABI 310, it was noted that the situation occurred only with some allele 8s, 9s, and 10s (see underlined alleles in the ABI 310 column of exhibit 18). Thus, these null alleles were variants of alleles with 8, 9, or

Exhibit 40. CDOJ DYS391 STR results with the mass spectrometry method

Position	Mass Spec	Allele 1 (Da)	Position	Mass Spec	Allele 1 (Da)
A1	10	24,489	A7	10	24,416
B1	11	25,672	B7	10	24,406
C1	10	24,487	C7	10	24,353
D1	10	24,455	D7	8	21,905
E1	10	24,471	E7	11	25,662
F1	11	25,641	F7	10	24,353
G1	11	25,637	G7	10	24,422
H1	10	24,436	H7	10	24,451
A2	10	24,459	A8	10	24,455
B2	10	24,465	B8	11	25,529
C2	10	24,440	C8	10	24,410
D2	10	24,455	D8	10	24,473
E2	10	24,359	E8	12	26,841
F2	11	25,837	F8	10	24,438
G2	10	24,444	G8	10	24,359
H2	10	24,451	H8	12	26,805
A3	11	25,639	A9	10	24,367
B3	10	24,414	B9	10	24,416
C3	11	25,654	C9	10	24,463
D3	11	25,591	D9	10	24,343
E3	10	24,463	E9	10	24,471
F3	11	25,648	F9	10	24,457
G3	10	24,475	G9	10	24,465
H3	10	24,436	H9	10	24,479
A4	10	24,408	A10	11	25,631
B4	10	24,475	B10	10	24,617
C4	11	25,625	C10	10	24,560
D4	No data		D10	10	24,444
E4	11	25,581	E10	11	25,650
F4	10	24,396	F10	10	24,414
G4	11	25,562	G10	10	24,453
H4	10	24,463	H10	11	25,866
A5	10	24,390	A11	11	25,662
B5	10	24,446	B11	10	24,432
C5	10	24,599	C11	10	24,400
D5	11	25,652	D11	10	24,463
E5	10	24,451	E11	10	24,420
F5	10	24,436	F11	9	23,175
G5	10	24,380	G11	10	24,459
H5	12	26,775	H11	11	25,585
A6	11	25,550			
B6	10	24,428			
C6	11	25,583			
D6	10	24,457			
E6	11	25,658			
F6	10	24,436			
G6	10	24,473			
H6	10	24,463			

Exhibit 41. **Plot of measured masses versus sample number from four different STR loci.** FGA, which has a high degree of scatter, is the most polymorphic marker and has the largest mass alleles. The highest and lowest alleles observed for each STR locus in this study are shown on the plot.

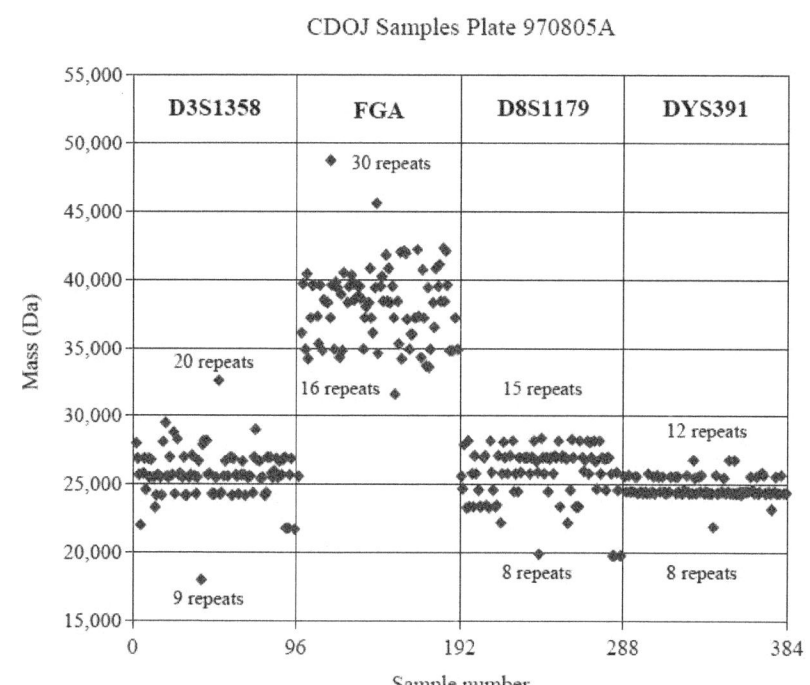

The ability to make accurate mass measurements with mass spectrometry is a potential advantage when locating new microvariants. If the mass precision is good, then any peaks that have large offsets from the expected full repeat alleles could be suspect microvariants in the form of insertions or deletions because their masses would fall outside the expected variance due to instrument variation. This possibility is especially true when working with heterozygous samples. Microvariants can be detected by using the mass difference between the two alleles and comparing this value with the expected value for full repeats or with the allele peak mass offsets. If the peak mass offsets shift together, then both alleles are full repeats, but if one of the peak mass offsets is significantly different (e.g., ~300 Da), a possible insertion or deletion exists in one of the alleles. Exhibit 46 illustrates this concept by plotting the mass offset (from a calculated allele mass) of allele 1 verses the mass offset (from a calculated allele mass) of allele 2. Note that the 9.3 microvariant (i.e., partial) repeat alleles for TH01 cluster away from the comparison of full repeat versus full repeat allele. On the other hand, results for the other three STR loci, which have no known microvariants in this data set, have mass offsets that shift together for the heterozygous alleles. Exhibit 47 compares the peak mass offsets for the amelogenin X allele with the Y allele and demonstrates that full "repeats" shift together during mass spectrometry measurements.

Nontemplate addition

DNA polymerases, particularly the *Taq* polymerase used in PCR, often add an extra nucleotide to the 3'-end of a PCR product as template strands are copied. This nontemplate addition—which is most often an adenine, hence, the term "adenylation"—can be favored by adding a final incubation step at 60 °C or 72 °C after the

10 repeats. Most likely, a sequence microvariant occurs within the repeat region near the 3'-end of the reverse primer, which anneals to two full repeats. Unfortunately, time constraints restricted the gathering of sequence information for these samples to confirm the observed variation. Interestingly enough, the D7S820 locus has been reported to cause similar null allele problems with other primer sets (Schumm et al., 1997).

Microvariants

Sequence variation between alleles can take the form of insertions, deletions, or nucleotide changes. Alleles containing some form of sequence variation compared with more commonly observed alleles are often referred to as microvariants because they are slightly different from full

repeat alleles. For example, the STR locus TH01 contains a 9.3 allele, which has 9 full repeats (AATG) and a partial repeat of 3 bases (ATG). In this particular example, the 9.3 allele differs from the 10 allele by a single base deletion of adenine. Microvariants exist for most STR loci and are being identified in greater numbers as more samples are being examined around the world. In this study, three previously unreported STR microvariants (exhibit 44) were discovered during the analysis of 38 genomic DNA samples from a male population data set provided by Dr. Oefner (exhibit 45). These microvariants occurred in the three most polymorphic STR loci that possess the largest and most complex repeat structures: FGA, D21S11, and D18S51.

temperature cycling steps in PCR (Clark, 1988, and Kimpton et al., 1993). However, the degree of adenylation is dependent on the sequence of the template strand, which in the case of PCR results from the 5'-end of the reverse primer. Thus, every locus will have different adenylation properties because the primer sequences are different. From a measurement standpoint, it is better to have all molecules of a PCR product as similar as possible for a particular allele. Partial adenylation, where some of the PCR products do not have the extra adenine (i.e., -A peaks) and some do (i.e., +A peaks), can contribute to peak broadness if the separation system's resolution is poor (see top panel of exhibit 37). Sharper peaks improve the likelihood that a system's genotyping software can make accurate calls. Variation in the adenylation status of an allele across multiple samples can have an impact on accurate sizing and genotyping potential microvariants. For example, a nonadenylated TH01 10 allele would look the same as a fully adenylated TH01 9.3 allele in the mass spectrometer because their masses are identical. Therefore, it is beneficial if all PCR products for a particular amplification are either +A or -A rather than a mixture (e.g., ±A). By using the temperature soak at the end of thermal cycling, most of the STR loci were fully adenylated, with the notable exception of TPOX, which was typically nonadenylated, and TH01, which under some PCR conditions produced partially adenylated amplicons. For making correct genotype calls, the STR mass ladder file (exhibit 30) was altered according to the empirically determined adenylation status.

During the course of this project, Platinum® GenoTYPE™ *Tsp* DNA polymerase (Life Technologies, Rockville, MD) became available that exhibits little to no nontemplate nucleotide addition. This new DNA polymerase was tested with STR loci that had been shown to produce partial adenylation to see if the +A peak could be eliminated. Exhibit 48 compares mass spectrometry results obtained using AmpliTaq Gold (commonly used) polymerase with the new *Tsp* polymerase. The *Tsp* polymerase produced amplicons with only the -A peaks, while TaqGold showed partial adenylation with these TH01 primers. Thus, this new polymerase has the potential to produce sharper peaks (i.e., no partial adenylation) and allele masses that can be more easily predicted

Exhibit 42. **Effects of sequence variation on PCR amplification in or around STR repeat regions.** The asterisk symbolizes a DNA difference (base change, insertion, or deletion of a nucleotide) from a typical allele for a STR locus. In situation (A), the variation occurs within the repeat region (depicted in green) and should have no impact on the primer binding and the subsequent PCR amplification, although the overall amplicon size may vary slightly. In situation (B), the sequence variation occurs just outside the repeat in the flanking region but interior to the primer annealing sites. Again, PCR should not be affected, although the size of the PCR product may vary slightly. However, in situation (C), the PCR can fail due to a disruption in annealing a primer because the primer no longer perfectly matches the DNA template sequence. Therefore, if sequence variation occurs in the flanking region for a particular locus, one set of primers may work while another may fail to amplify the template. The template would therefore be a "null" allele.

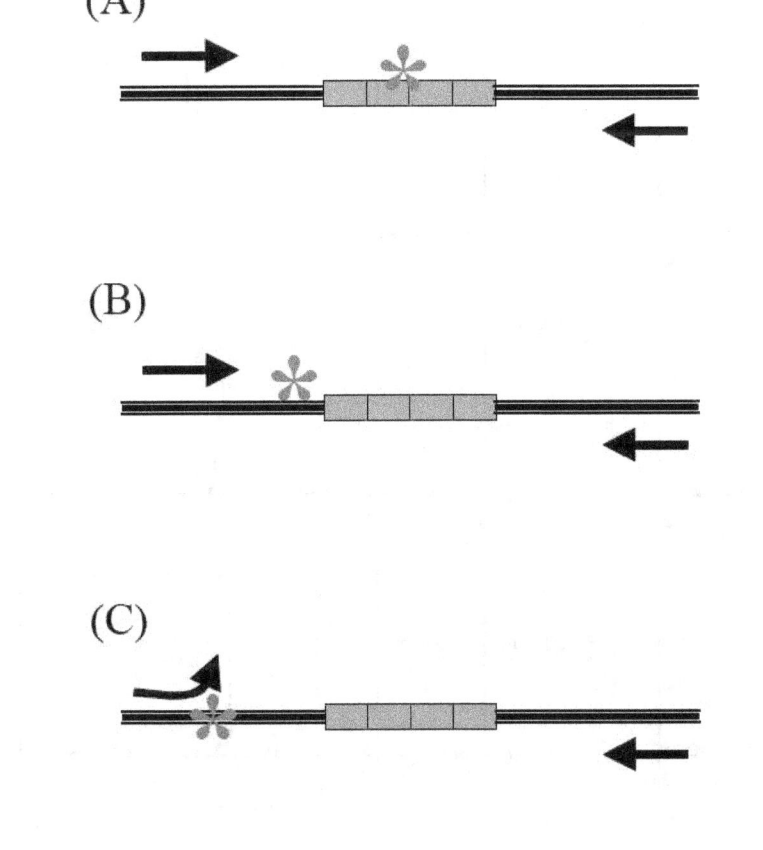

(A)

(B)

(C)

(i.e., all PCR products would be nonadenylated).

Stutter products

During PCR amplification of STR loci, repeat slippage can occur and result in the loss of a repeat unit as DNA strand synthesis occurs through a repeated sequence. These stutter products are typically 4 bases, or one tetranucleotide repeat, shorter than the true allele PCR product. The amount of stutter product compared to the allele product varies depending on the STR locus and the length of the repeat, but typically stutter peaks are 2–10% of the allele peak height (Walsh et al., 1996). Forensic DNA scientists are concerned about stutter products because their presence can interfere in the interpretation of DNA mixture profiles.

When reviewing plots of GeneTrace's mass spectrometry results for STR loci, forensic scientists have commented on the reduced level of stutter product

Exhibit 43. **Mass spectra of CDOJ samples amplified with D7S820 primers.** From left side, top-to-bottom followed by right side, top-to-bottom, sample genotypes are (10,11), (9,10), (10,10), (null allele 8, 10), (8,11), (11,13), (10,12), and (8,10). The arrow indicates the position where allele 8 should be present in the sample but is missing due to a primer annealing binding site sequence polymorphism that results in a "null allele." The mass range shown here is 12,000–25,000 Da.

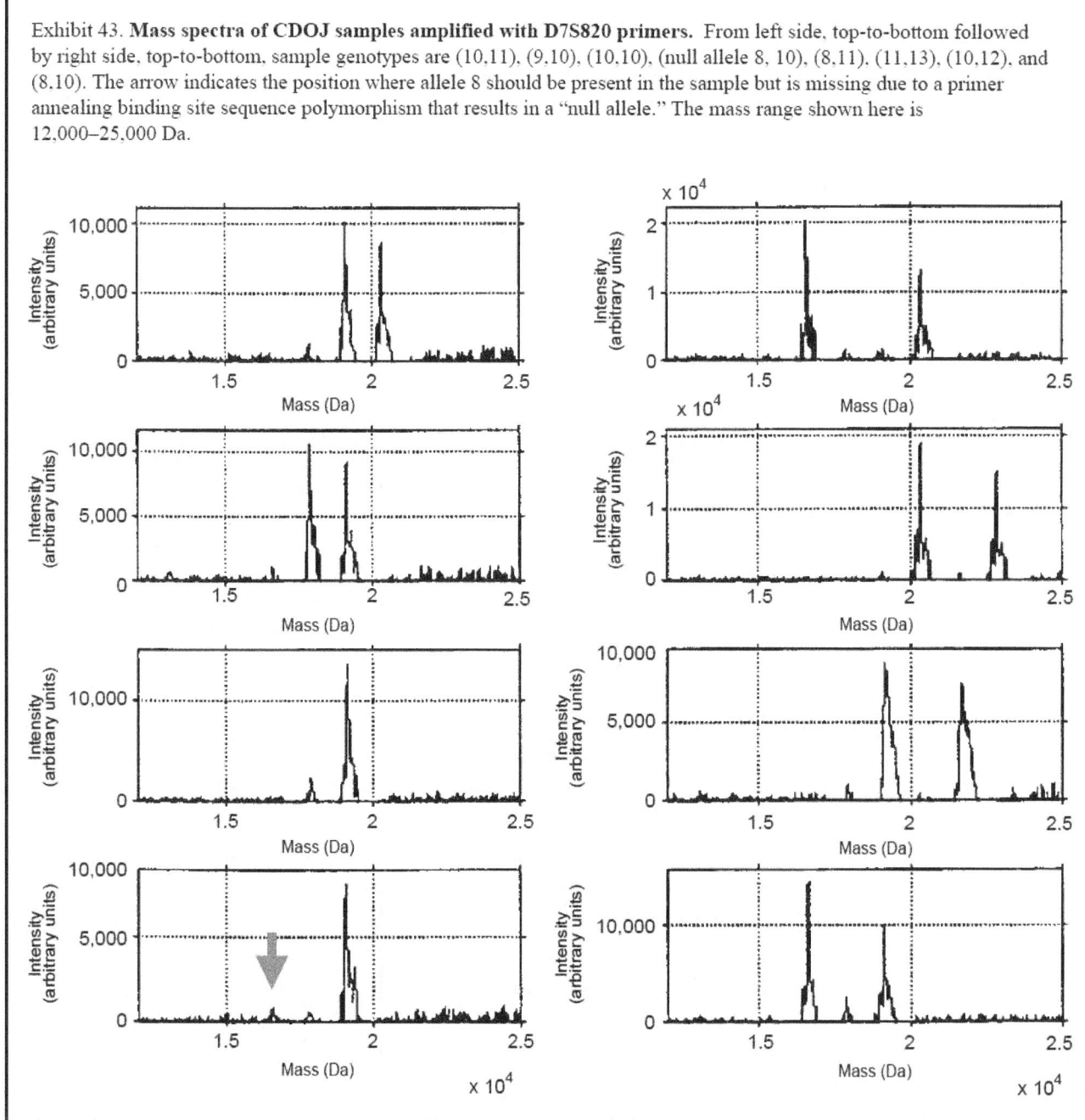

detection (exhibit 35). There are two possibilities for this reduction:

◆ Since the primers are closer to the repeat region, smaller PCR products are amplified, which means that the DNA polymerase does not have to hold on to the extending strand as long for synthesis purposes. It is possible that the polymerase reads through the repeat region "faster" and, therefore, the template strands do not have as much of an opportunity to slip and reanneal out of register on the repeat region. For example, *Taq* polymerase has a processivity rate of ~60 bases before it falls off the extending DNA strand; therefore, the closer the PCR product size is to 60 bases, the better the extension portion of the PCR cycle. GeneTrace's PCR product sizes, which are typically less than 100 bp, are much smaller than the

Exhibit 44. **Electropherograms of ABI 310 results for new STR microvariants seen in the Stanford male population samples.** The D18S51 16.2 allele, D21S11 30.3 allele, and FGA 28.1 allele have not been reported previously in the literature. These plots are views from Genotyper 2.0 with results overlaid on shaded allele bins. The base pair size range is indicated at the top of each plot. Note: The microvariant alleles (indicated by the red arrows) fall between the shaded bins, but the other alleles in the heterozygote set contain complete repeats and fall directly on the shaded (expected) allele bin.

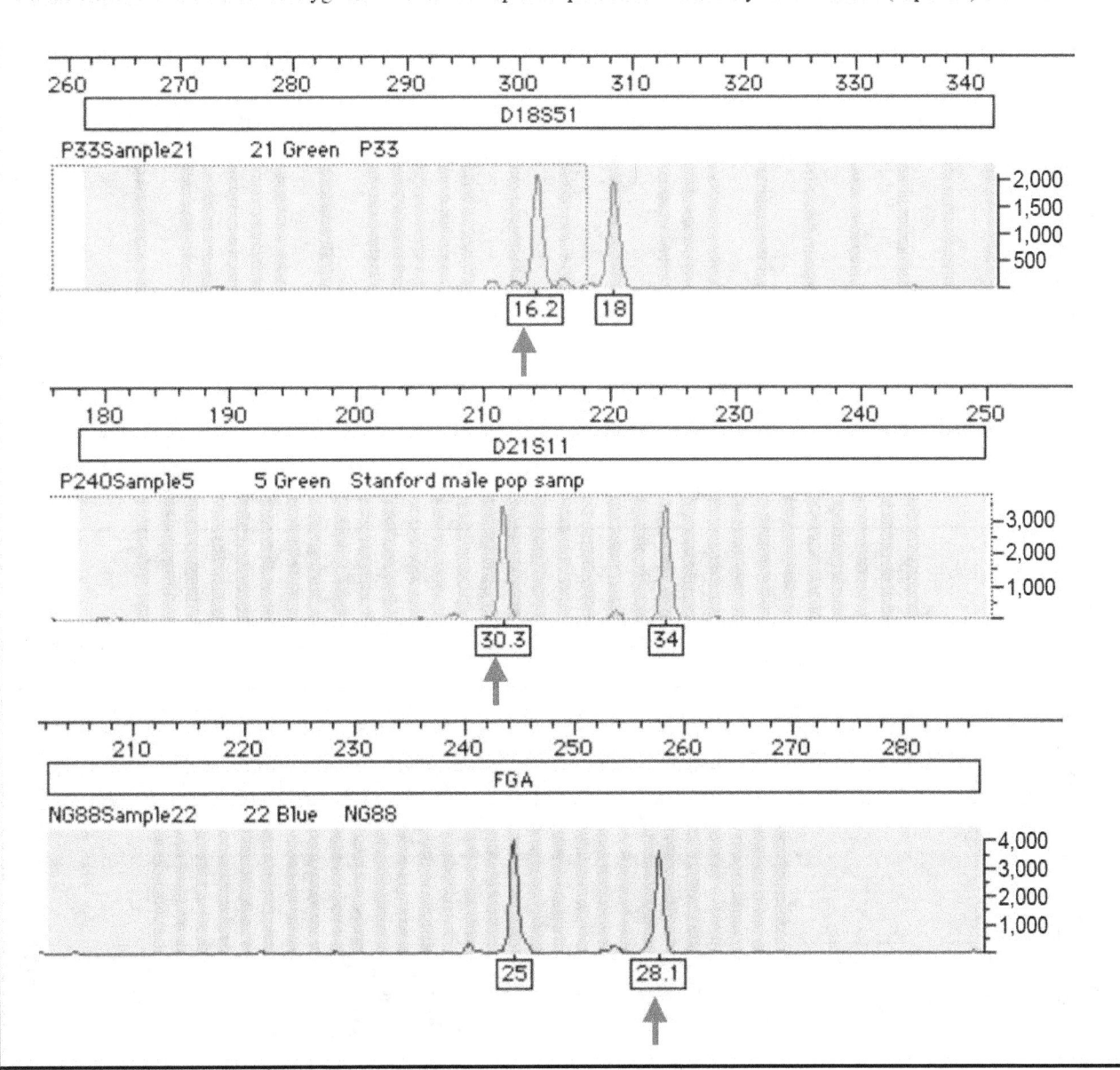

Exhibit 45. ProfilerPlus™ results from Stanford male population samples

Sample Name	Amel-A1	Amel-A2	D8-A1	D8-A2	D21-A1	D21-A2	D18-A1	D18-A2	D3-A1	D3-A2	VWA-A1	VWA-A2	FGA-A1	FGA-A2	D5-A1	D5-A2	D13-A1	D13-A2	D7-A1	D7-A2
Aus21	X	Y	8	12	29	31.2	12	18	15	17	14	17	18	22	12	12	11	12	9	11
Aus28	X	Y	14	15	29	31.2	15	19	16	17	15	17	19	20	12	12	10	12	8	10
Berg15	X	Y	10	12,13	28,30.2	31.2	12,17	14	15,17	18,19	16	17,18	20,21	24	11	13w	8,14	11	7w	10
Berg19	X	Y	14	14	32.2	32.2	12	18	16	16	14	18	22	22	12	13	12	12	9	13
Bl12	X	Y	12	12	30	31.2	17	17	16	18	17	19	24	27	12	12	11	11	9	11
Bsk092	X	Y	14	14	29	33.2	13	16	17	17	17	17	23	23	11	12	12	12	11	11
Bsk111	X	Y	12	13	30	32.2	16	19	14	14	16	16	18	19	12	12	8	10	8	9
Bsk118	X	Y	13	14	30.2	30.2	11	14	16	16	17	17	23	25	10	12	8	10	8	8
CH17	X	Y	12	14	32.2	34.2	14	23	15	15	16	18	20	26	9	12	8	12	11	12
CH23	X	Y	12	13	29	31	16	16	15	15	14	17	24	24	11	12	9	12	10	12
CH42	X	Y	12	14	28.2	31.2	13	15	16	16	17	17	21	22	12	12	10	12	11	11
F18	X	Y	14	14	31	32.2	19	19	15	15	14	16	21	23	13	13	11	12	8	12
F21	X	Y	14	15	27	31	14	19	15	17	17	18	22	26	13	13	11	13	8	8
J13	X	Y	14	14	32.2	33.2	17	19	16	16	14	16	20	22	10	11	8	11	10	10
J3	X	Y	13	14	30	30	13	15	15	16	17	17	22	24	10	14	8	11	10	10
J37	X	Y	10	13	31	31.2	13	20	15	15	16	18	22	25	10	12	8	13	10	10
J39	X	Y	13	14	30	31.2	13	13	15	18	17	17	23	24	11	13	12	11	8	12
JK2921	X	Y	10	10	29	31.2	14	15	15	16	14	19w	19	22	10	11	11	11	9	9
JK2979	X	Y	10	12	32.2	33.2	14	14	15	16	14	18	21	24.2	11	14	11	12	11	11
MDK204	X	Y	14	17	28	30.2	12	19	14	15	16	16	23	25	12	13	11	12	8	10
Mel12	X	Y	13	14	30	32.2	17	20	15	16	17	18	21	22	13	14,15	11	12	12	12
Mel15	X	Y	14	15	28	31.2	14,15	22	16	18	16	18	23	24	10	10	12	13	11	11
Mel18	X	Y	15	15	28	32.2	14	15	15	16	17	18	24	24	10	11	12	13	10	11
GN 83	X	Y	13	13	30	31	13	18	15	16	15	17	23	24	11	11	8	8	8	8
NG85	X	Y	15	15	29	31.2	15	21	16	17	17	19	23	26	13	13	8	8	9	10
NG88	X	Y	12	15	33.2	38.2	13	17	15	16	16	16	25	22	11	11	8	11	8	12
OM135	X	Y	12	13	28	39	15	16	16	16	16	18	21	25	12	12	12	12	11	10
P103G	X	X	12	17	29	30	19	20	17	17	15	19	20	24	13	13	10	12	8	8
P109	X	Y	11	14	28	34	17	21	15	16	15	19	22	27	11	12	12	12	9	10
P205	X	Y	11	15	28	36	15	15	14	15	15	16	21	28.1	8	10	12	12	9	10
P240	X	Y	14	14	30.3	34	16	16	16	17	15	21	22	22.2	10	11	8	13	11	12
P33	X	Y	13	14	29	30	16.2	18	14	17	16	17	20	24	8	10	11	12	10	10
P37G	X	Y	14	15	29	32.2	14	20	16	17	17	18	24	24	12	13	9	12	8	11
P37G?	X	Y	14	15	29	32.2	14	20	15	17	17	18	24	24	12	13	9	12	8	11
P73	X	Y	14	15	29	30	14	16	15	16	14	19	22.2	24	12	12	11	13	10	11
PG1162	X	Y	8	14	31.2	32.2	14	15	16	16	16	18	18	23	11	12	9	11	8	10
PKH062	X	Y	12	15	29	31	13	16	16	18	16	18	21	22	11	11	9	13	10	12
SDH053	X	Y	10	12	30	33.2	16	16	18	18	17	17	21	25	11	12	8	8	8	8

Note: The STR loci are color coded to indicate their fluorescent dye label color; the shaded boxes in the body of the exhibit refer to the microvariant alleles (exhibit 44), 3-banded patterns, or an unexpected "x,x" amelogenin.

Exhibit 46. **Plot of allele mass offsets (allele 1 versus allele 2) for heterozygous samples from four different loci.** These are 88 CDOJ samples for the STR loci TH01, TPOX, CSF1PO, and D16S539.

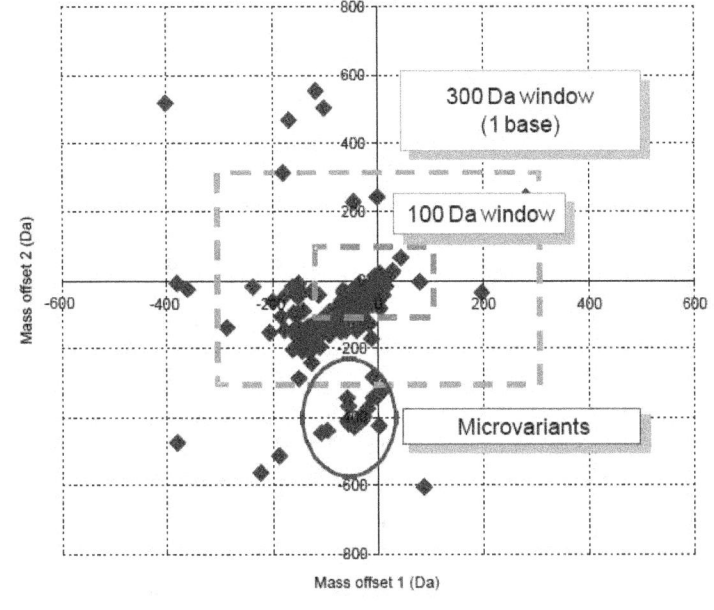

Exhibit 47. **Plot of X allele mass offset versus Y allele mass offset for 88 amelogenin samples.** The red box shows ±100 Da around the expected values and the green box shows ±300 Da. The blue line is the ideal situation where "heterozygous" peaks would shift in unison compared with the expected masses.

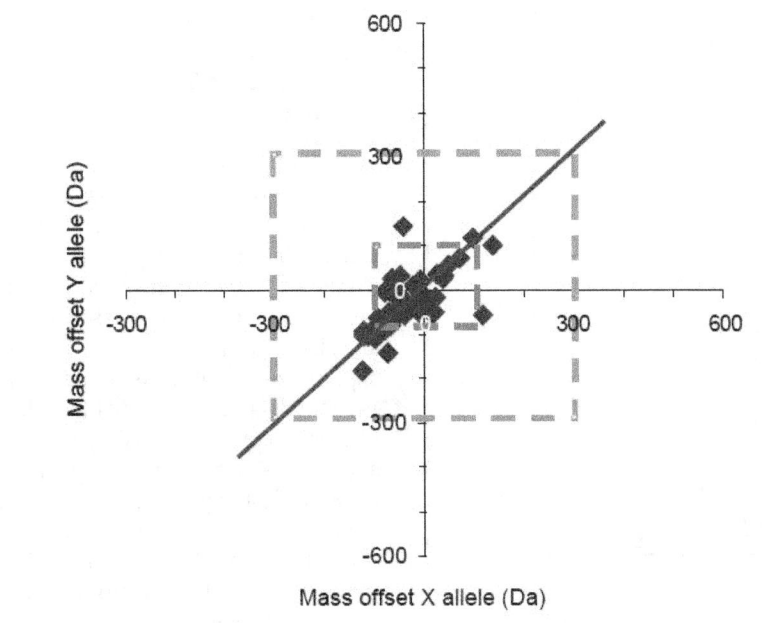

fluorescently labeled primer sets used by most forensic DNA laboratories (exhibit 2). However, this needs to be studied more extensively with multiple primer sets on a particular STR locus that generates various sized amplicons. For example, the primer sets described in exhibit 24 could be fluorescently labeled and analyzed on the ABI 310, where the stutter product peak heights could be quantitatively compared to the allele peak heights.

◆ The more likely reason that less stutter is observed by mass spectrometry is that the signal-to-noise ratio is much lower in mass spectrometry than in fluorescence measurements. Fluorescence techniques have a much lower background and are more sensitive for the detection of DNA than mass spectrometry. Thus, stutter may be present at similar ratios compared with those observed in fluorescence measurements, but because stutter is part of the baseline noise of mass spectrometry data, it may not be seen in the mass spectrum. This latter explanation is probably more likely, as indicated in very strong stutter peaks for some dinucleotide repeat markers (exhibit 49).

Whether stutter products are present or not, GeneTrace's current STR genotyping software has been designed to recognize them and not call them as alleles.

Primer sequence determinations from commercial STR kits

Primarily, two commercial manufacturers supply STR kits to the forensic DNA community: Promega Corporation and Applied Biosystems. These kits come with PCR primer sequences that permit simultaneous multiplex PCR amplification of up to 16 STR loci. One of the primers for each STR locus is labeled with a fluorescent

Exhibit 48. **Mass spectra comparing an STR sample amplified with TaqGold polymerase and Tsp polymerase.** The Tsp polymerase favors production of the nonadenylated form of PCR products, which results in a single peak for each allele (bottom panel). TaqGold produces a mixture of –A and +A peaks, which leads to two peaks for each allele (top panel). The peak masses in Daltons are indicated next to each peak. Mass difference measurements of 308 Da and 305 Da between the +A and –A peaks reveal that a "T" is added by TaqGold instead of the expected "A" (expected masses: T = 304 Da and A = 313 Da). The sample's genotype was TH01 6,8.

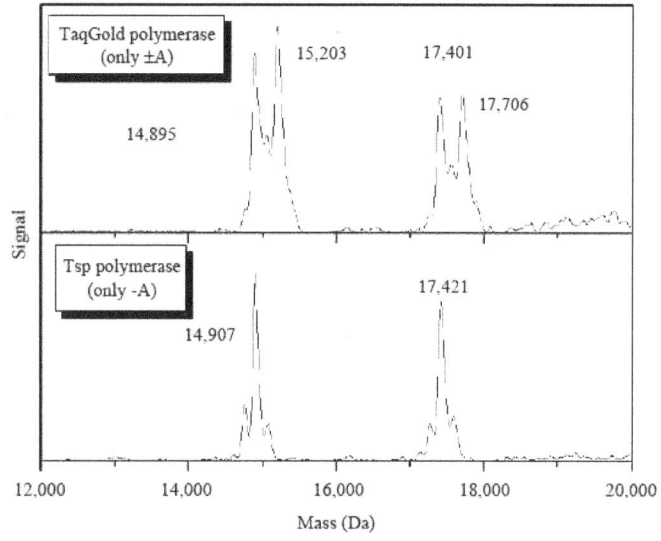

Exhibit 49. **Mass spectrum demonstrating detection of stutter products from a particularly stutter-prone dinucleotide repeat locus.** The mass differences between the stutter product peaks and the allele peaks can be used to determine the repeat sequence that is present on the measured DNA strand. Note: The amount of stutter is larger in the longer repeat allele than in the shorter allele.

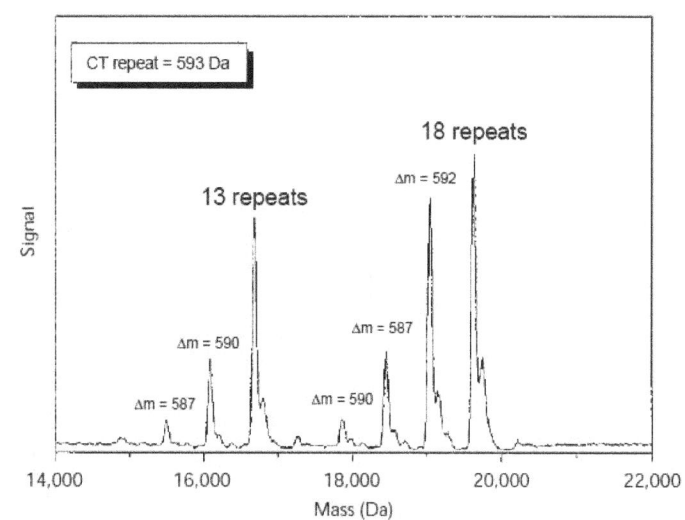

dye to permit fluorescent detection of the labeled PCR products. Since the primer sequences are not disclosed by the manufacturers, mass spectrometry was used to determine where they annealed to the STR sequences compared with GeneTrace primers (see previous discussion on null alleles).

First, the primer mixtures were spotted and analyzed to determine each primer's mass (top panel of exhibit 50). Then a 5'→3' exonuclease was added to the primer mix and heated to 37 °C for several minutes to digest the primer one base at a time. An aliquot was removed every 5–10 minutes to obtain a time course on the digestion reaction. Each aliquot was spotted in 3-hydroxypicolinic acid matrix solution (Wu et al., 1993), allowed to dry, and analyzed in the mass spectrometer.

A digestion reaction produces a series of products that differ by one nucleotide. By measuring the mass difference between each peak, the original primer sequence may be determined (bottom panel of exhibit 50). Only the unlabeled primers will be digested because the covalently attached fluorescent dye blocks the 5'-end of the dye-labeled primer. Using only a few bases of sequence (e.g., 4–5 bases), it is possible to make a match on the appropriate STR sequence obtained from GenBank to determine the 5'-end of the primer without the fluorescent label.

With the full-length primer mass obtained from the first experiment, the remainder of the unlabeled primer can be identified. The position of the 5'-end of the other primer can be determined using the GenBank sequence and the PCR product length for the appropriate STR allele listed in GenBank (exhibit 2). The sequence of the labeled primer can be ascertained by using the appropriate primer mass determined from the first experiment and subtracting the mass of the fluorescent dye. The primer mass is then used to obtain the correct length of the primer on the GenBank sequence and the primer's

Exhibit 50. **Primer sequence determination with exonuclease digestion and mass difference measurements.** This example is a D5S818 primer pair purchased from Promega Corporation and used in its PowerPlex™ STR kit. The top panel shows a mass spectrum of the original primer pair prior to digestion. The bottom panel is the mass spectrum of the same primers following a 6 minute digestion at 37 °C with calf spleen phosphodiesterase, which is a 5'→3' exonuclease. The dye-labeled primer is not digested because the dye protects the 5'-end of the primer. Mass difference measurements between the digestion peaks leads to the sequence determination of the 5'-end of the unlabeled primer (see underlined portion of forward sequence). The determined sequences are 5'-GGTGATTTTCCTCTTTGGTATCC-3' (forward) and 5'-fluorescein dye—TTTACAACATTTGTATCTATATCTGT-3' (reverse).

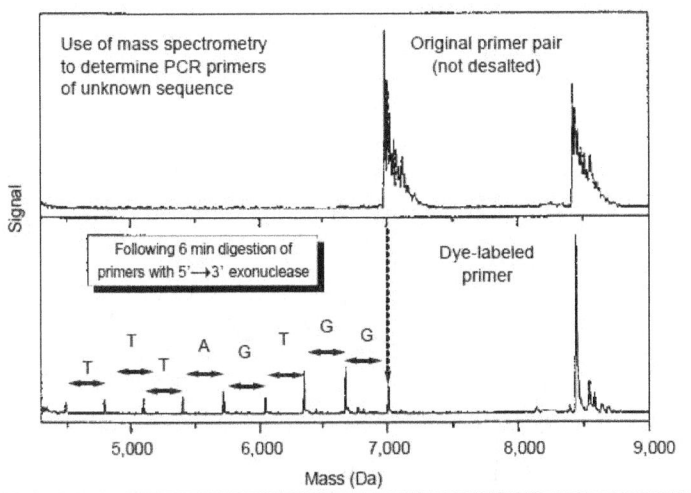

Exhibit 51. **Mass spectrum of AmpFlSTR® Green I primer mix.** Each peak has been identified with its corresponding primer. Peaks containing the fluorescent dye (JOE) are underlined.

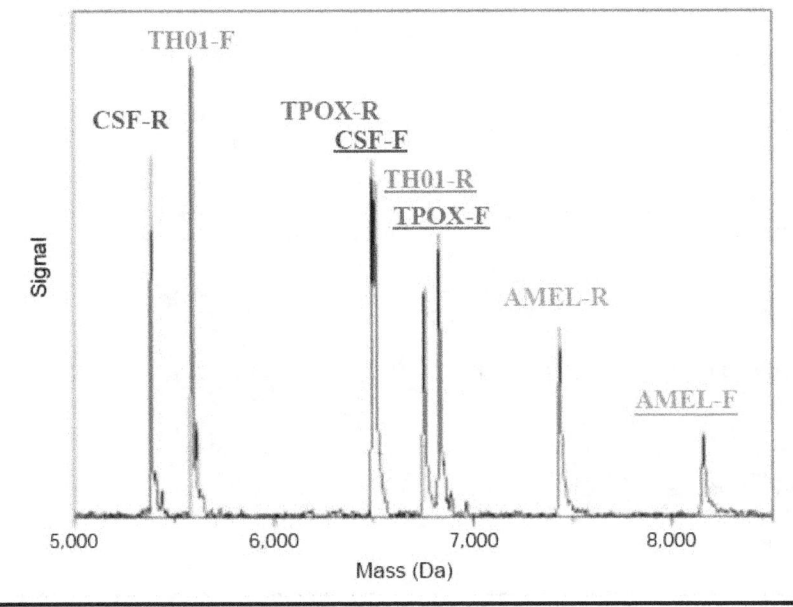

sequence. Finally, an entire STR multiplex primer set can be measured together in the mass spectrometer to observe the primer balance (exhibit 51). High-performance liquid chromatography fraction collection can be used to pull primers apart from complex, multiplex mixtures, and each primer can be identified as previously described. The primer sequences from both Promega and Applied Biosystems STR loci TH01 (exhibit 52), TPOX (exhibit 53), and CSF1PO (exhibit 54) were identified using this procedure. A comparison of the primer sequences from the two manufacturers found that they were very similar. The 3'-ends of the primer sets—the most critical portions for annealing during—were almost identical between the different kits. The ABI primers were typically shorter at the 5'-end and, therefore, produced PCR products that were ~10 bases shorter than those produced by the corresponding Promega primers. In all three STR loci, the primers annealed further away from the repeat region than the GeneTrace primer sets.

Analytical Capabilities of This Mass Spectrometry Method

Using the current primer design strategy, most STR alleles ranged in size from ~10,000 Da to ~40,000 Da. In mass spectrometry, the smaller the molecule, the easier it is to ionize and detect (all other things being equal). Resolution, sensitivity, and accuracy are usually better the smaller the DNA molecule being measured. Because the possible STR alleles are relatively far apart, reliable genotyping is readily attainable even with DNA molecules at the higher mass region of the spectrum. For example, neighboring full-length alleles for a tetranucleotide repeat, such as AATG, differ in mass by 1,260 Da.

Exhibit 52. **TH01 STR primer positions for commercially available primers highlighted on the GenBank sequence.** The forward primer is shown in blue with the reverse primer in brown. The repeat is highlighted in green on the strand that contains the fluorescent dye. The 3'-positions of the forward primers are identical but differ by a single base for the reverse primers. Promega primers are longer at the 5'-end, which produces a larger PCR product by 11 bp (6 bases on forward and 5 bases on reverse).

AmpFlSTR® Green I Kit

Reverse primer is labeled with JOE dye (fluorescein derivative)

PCR product =184 bp (9 repeats)

PowerPlex™ Kit

Forward primer is labeled with TMR dye (tetramethylrhodamine)

PCR product = 195 bp (9 repeats)

Exhibit 53. **TPOX STR primer positions for commercially available primers highlighted on the GenBank sequence.** The forward primer is shown in blue with the reverse primer in brown. The repeat is highlighted in green on the strand that contains the fluorescent dye. The 3'-positions of the reverse primers are identical but differ by a single base for the forward primers. Promega primers are longer at the 5'-end, which produces a larger PCR product by 7 bp (4 bases on forward and 3 bases on reverse).

AmpFlSTR® Green I Kit

Forward primer is labeled with JOE dye (fluorescein derivative)

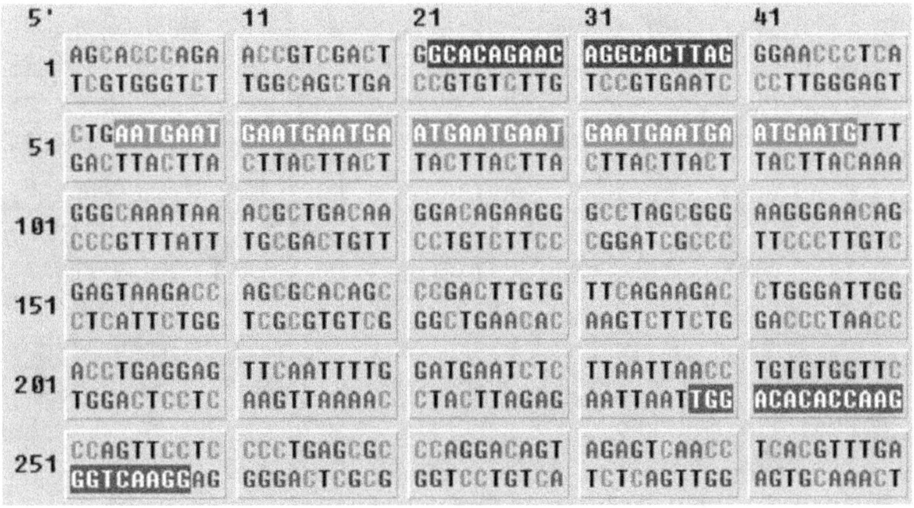

PCR product = 237 bp (11 repeats)

PowerPlex™ Kit

Reverse primer is labeled with TMR dye (tetramethylrhodamine)

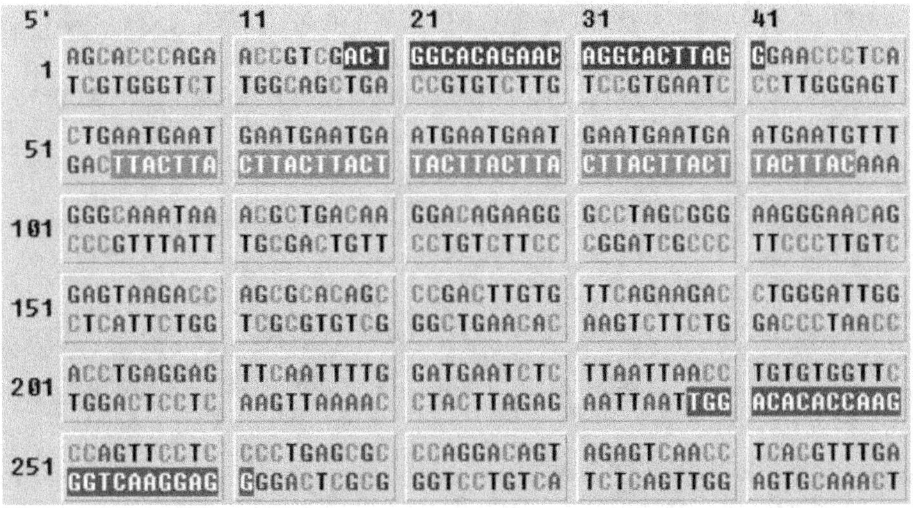

PCR product = 244 bp (11 repeats)

Exhibit 54. **CSF1PO STR primer positions for commercially available primers highlighted on the GenBank sequence.** The forward primer is shown in blue with the reverse primer in brown. The repeat is highlighted in green on the strand that contains the fluorescent dye. The 3'-positions of the reverse primers are identical but differ by a single base for the forward primers. Promega primers are longer at the 5'-end, which produces a larger PCR product by 11 bp (5 bases on forward and 6 bases on reverse).

AmpFlSTR® Green I Kit
Forward primer is labeled with JOE dye (fluorescein derivative)

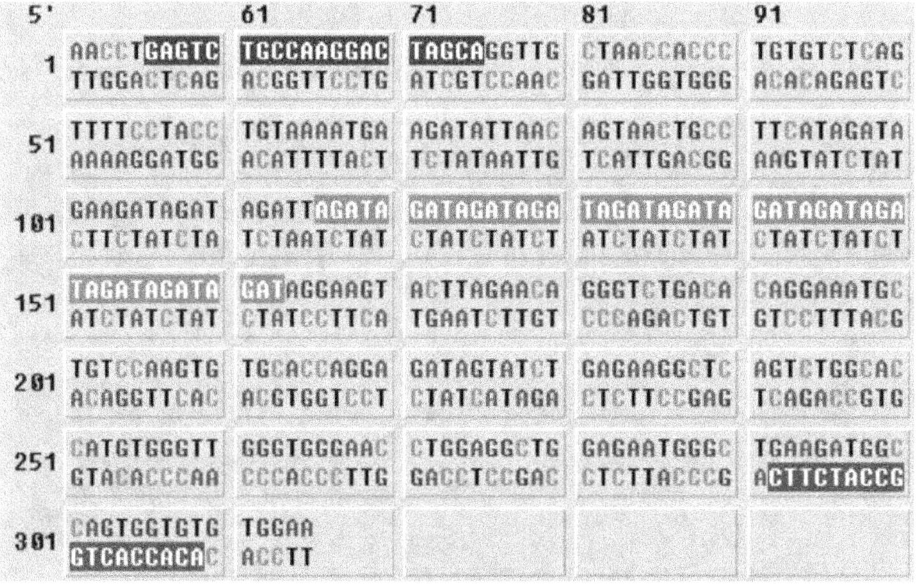

PCR product = 304 bp (12 repeats)

PowerPlex™ Kit
Forward primer is labeled with TMR dye (tetramethylrhodamine)

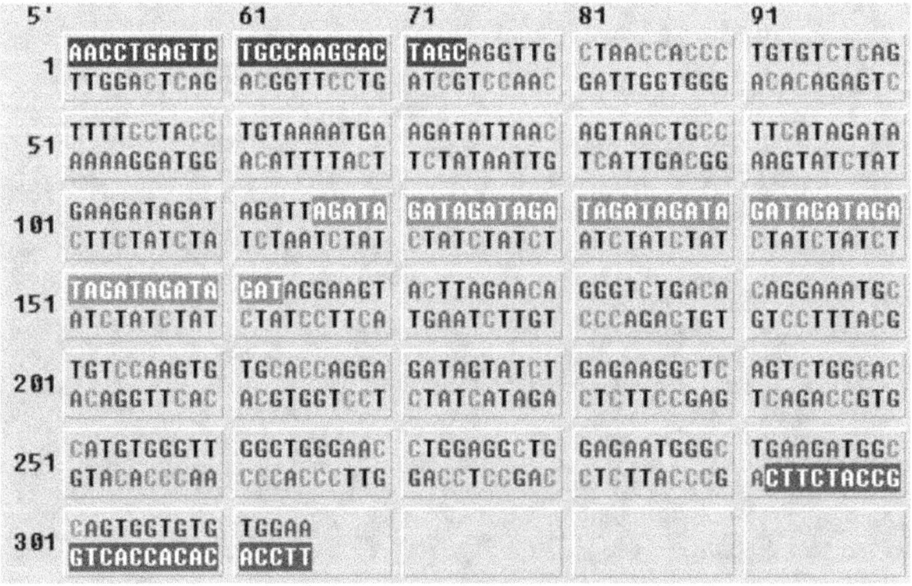

PCR product = 315 bp (12 repeats)

Exhibit 55. **Mass spectrum of a TH01 allelic ladder reamplified from AmpF1STR® Green I allelic ladders.** The PCR product size of allele 10 is only 83 bp with a measured mass of 20,280 Da and separation time of 204 μs. The allele 9.3 and allele 10 peaks, which are only a single nucleotide apart, differ by only 1.5 μs on a separation timescale and can be fully resolved with this method.

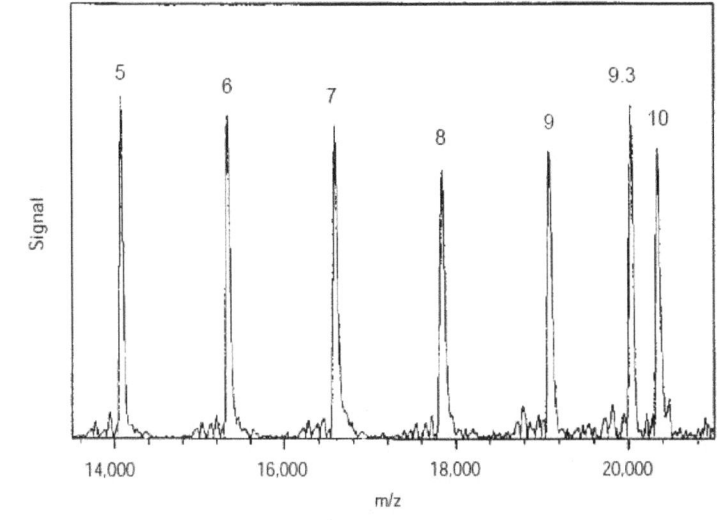

Exhibit 56. **Mass spectra of D5S818 allelic ladders from two manufacturers.** The Applied Biosystems D5S818 ladder contains 10 alleles (top panel); the Promega D5S818 ladder contains only 8 alleles (bottom panel). The GeneTrace primers bind internally to commercially available multiplex primers, and all alleles in the commercial allelic ladders are therefore amplified, demonstrating that the GeneTrace primers can amplify all common alleles for this particular STR locus.

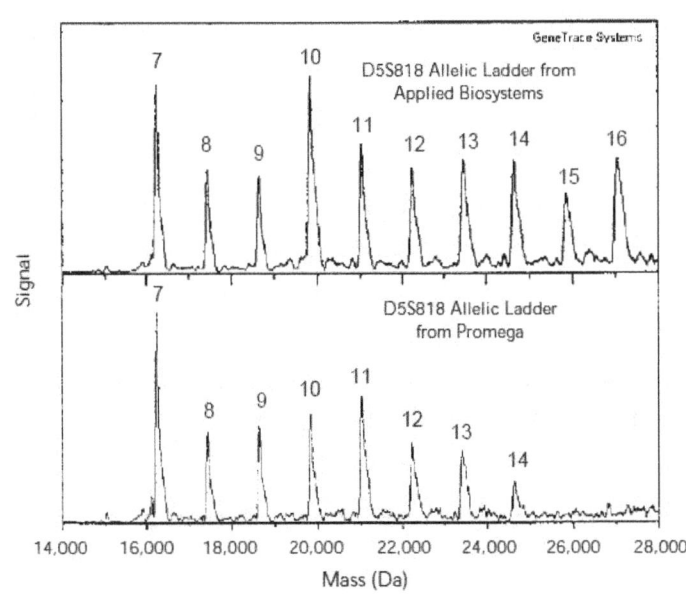

Resolution

Dinucleotide repeats, such as CA repeats, require a resolution of at least 2 bp in order to resolve stutter products from the true allele or heterozygotes that differ by a single repeat. Trinucleotide and tetranucleotide repeats, with their larger repeat structure, are more easily resolved because there is a larger mass difference between adjacent alleles. However, the overall mass of the PCR product increases more rapidly with tri- or tetranucleotide repeats. For example, the repeat region for 40 GA repeats is 25,680 Da, while the mass of the repeat region quickly increases to 37,200 Da for 40 AAT repeats and 50,400 Da for 40 AATG repeats.

GeneTrace has demonstrated that a resolution of a single dinucleotide repeat (~600 Da) may be obtained for DNA molecules up to a mass of ~35,000 Da. This reduced resolution at higher mass presents a problem for polymorphic STR loci such as D18S51, D21S11, and FGA because single base resolution is often required to accurately call closely spaced alleles or to distinguish a microvariant containing a partial repeat from a full-length allele. These three STR loci also contain long alleles. For example, D21S11 has reported alleles of up to 38 repeats (mixture of TCTA and TCTG) in length, D18S51 up to 27 AGAA repeats, and FGA up to 50 repeats (mixture of CTTT and CTTC). Heterozygous FGA alleles that differed by only a single repeat were more difficult to genotype accurately than smaller sized STR loci due to poor resolution at masses greater than ~35,000 Da (see samples marked in red in exhibit 19).

The analysis of STR allelic ladders demonstrates that all alleles can be resolved for an STR locus. Allelic ladders from commercial kits were typically diluted 1:1000 with deionized water and then reamplified with the

GeneTrace primers that bound closer to the repeat region than the primers from the commercial kits. This reamplification provided PCR products for demonstrating that the needed level of resolution (i.e., distinguishing adjacent alleles) is capable at the appropriate mass range in the mass spectrometer as well as demonstrating that the GeneTrace primers amplify all alleles (i.e., no allele dropout from a null allele). A number of STR allelic ladders were tested in this fashion, including TH01 (exhibit 55); CSF1PO, TPOX, and VWA (exhibit 36); and D5S818 (exhibit 56). All tetranucleotide repeat alleles were resolvable in these examples, demonstrating 4 bp resolution, and TH01 single base pair resolution was seen between alleles 9.3 and 10.

Sensitivity

To determine the sensitivity of Gene-Trace's STR typing assay, TPOX primers were tested with a dilution series of K562 genomic DNA (20 ng, 10 ng, 5 ng, 2 ng, 1 ng, 0.5 ng, 0.2 ng, and 0 ng). Promega's Taq polymerase and STR buffer were used with 35 PCR cycles as described in the scope and methodology section. Peaks for the correct genotype (heterozygote 8,9) could be seen down to the lowest level tested (0.2 ng or 200 picograms), while the negative control was blank. Exhibit 57 contains a plot with the mass spectra for 20 ng, 5 ng, 0.5 ng, and 0 ng. While each PCR primer pair can exhibit a slightly different efficiency, human DNA down to a level of ~1 ng can be reliably PCR amplified and detected using mass spectrometry. GeneTrace's most recent protocol involved 40-cycle PCR and the use of TaqGold™ DNA polymerase, which should improve overall yield for STR amplicons. All of the samples tested from CDOJ were amplified with only 5 ng of DNA template and yielded excellent results (exhibits 12–19, 31, 39, 40, 43, 58, and 59). In terms of absolute sensitivity in the mass spec-

trometer, several hundred femtomoles of relatively salt-free DNA molecules were typically found necessary for detection. GeneTrace's PCR amplifications normally produced several picomoles of PCR product, approximately an order of magnitude more material than is actually needed for detection.

Mass accuracy and precision

Mass accuracy is an important issue for this mass spectrometry approach to STR genotyping, as a measured mass for a particular allele is compared with an ideal mass for that allele. Due to

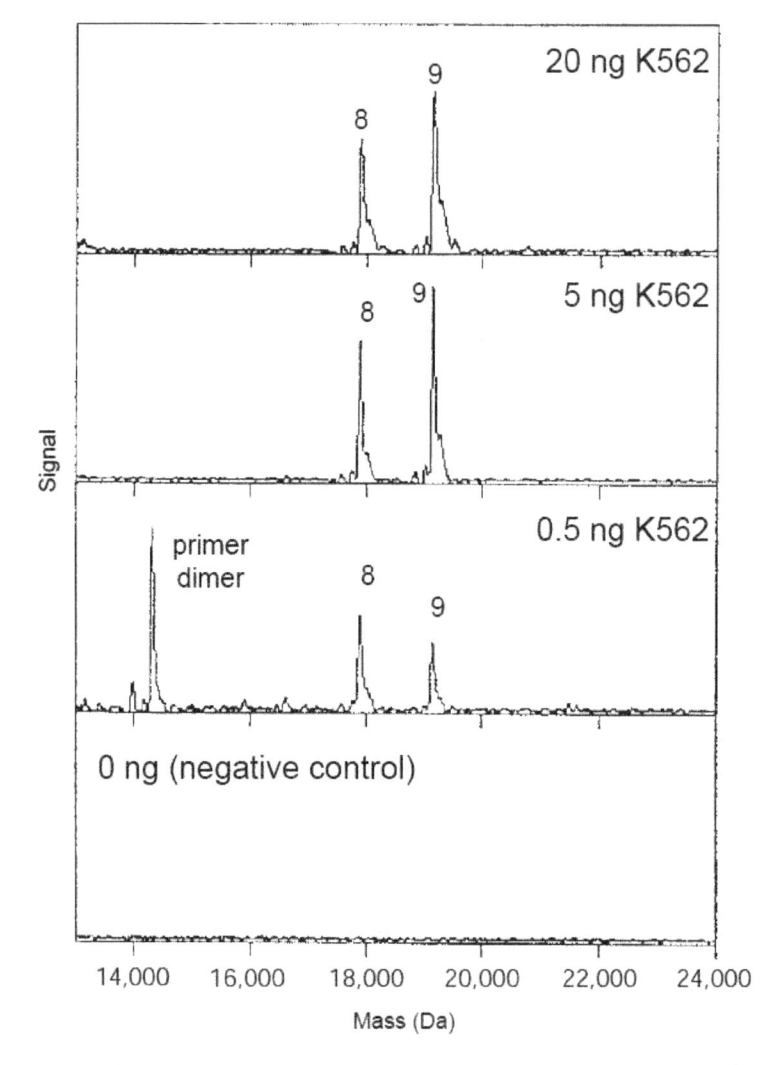

Exhibit 57. **Mass spectra of TPOX PCR products from various amounts of K562 DNA template material.** This sensitivity test demonstrates that DNA templates in the quantity range of 0.5–20 ng may be effectively amplified and detected by mass spectrometry.

the excellent accuracy of mass spectrometry, internal standards are not required to obtain accurate DNA sizing results as in gel or CE measurements (Butler et al., 1998). To make an inaccurate genotype call for a tetranucleotide repeat, the mass offset from an expected allele mass would have to be larger than 600 Da (half the mass of a ~1,200 Da repeat).

GeneTrace has observed mass accuracies on the order of 0.01 nucleotides (<3 Da) for STR allele measurements. However, under routine operation with GeneTrace's automated mass spectrometers, some resolution, sensitivity, and accuracy may be sacrificed compared with a research-grade instrument to deliver data at a high rate of speed. Almost all STR allele size measurements should be within ±200 Da, or a fraction of a single nucleotide, of the expected mass. Exhibit 60 illustrates that the precision and accuracy for STR measurements is good enough to make accurate genotyping calls with only a routine mass calibration, even when comparing data from the same samples collected months apart.

Exhibit 58. **Mass spectra of CDOJ samples amplified with TPOX primers.** From left side, top-to-bottom, followed by right side, top-to-bottom, sample genotypes are (8,9), (9,10), (10,11), (11,12), (8,8), (8,11), (11,11), and allelic ladder. The mass range shown here is 15,000–25,000 Da.

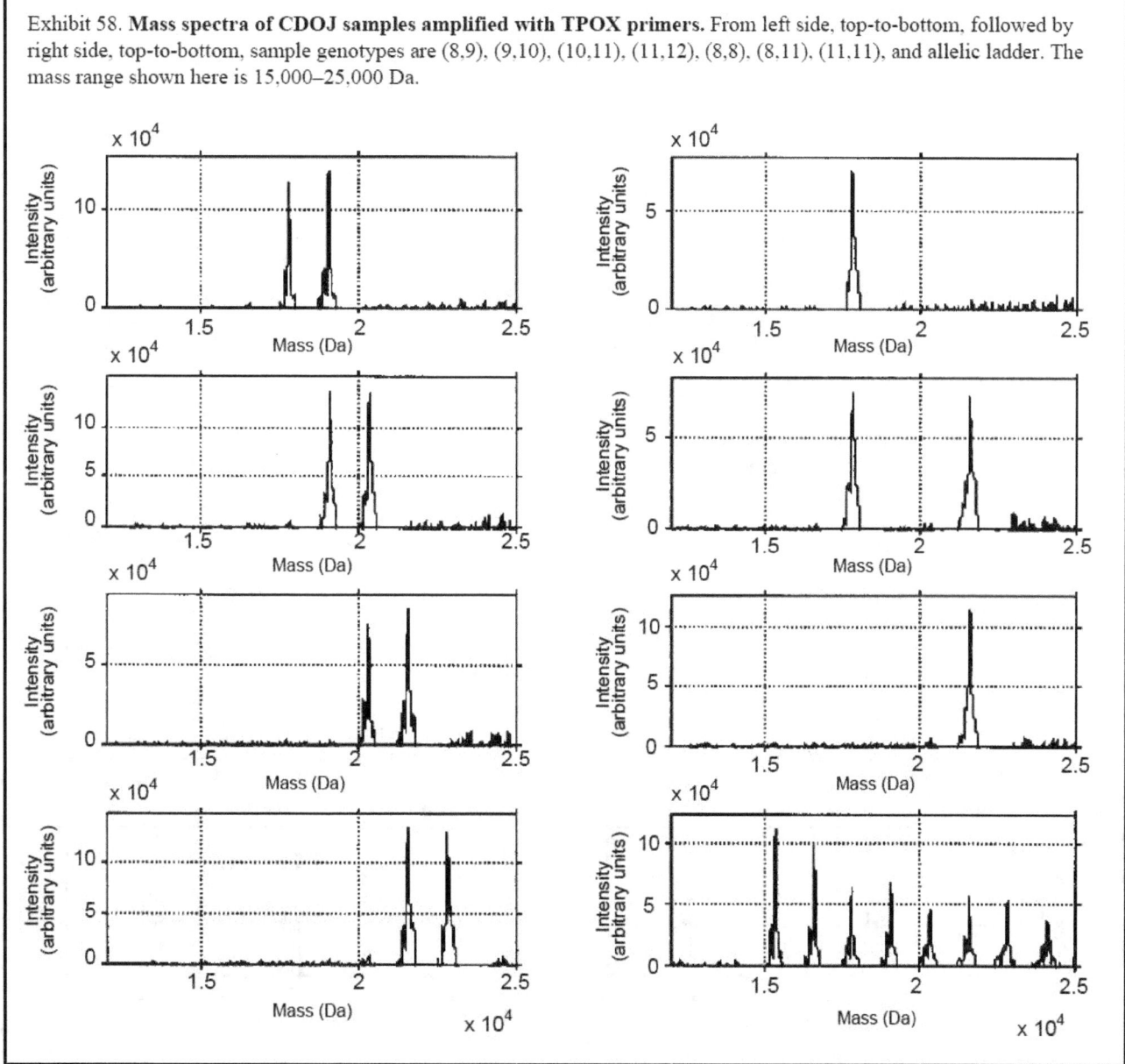

Precision is important for STR allele measurements in mass spectrometry because no internal standards are being run with each sample to make adjustments for slight variations in instrument conditions between runs. To demonstrate the excellent reproducibility of mass spectrometry, 15 mass spectra of a TPOX allelic ladder were collected. A table of the obtained masses for alleles 6, 7, 8, 9, 10, 11, 12, and 13 shows that all alleles were easily segregated and distinguishable (exhibit 61). Statistical analysis of the data found that the standard deviation about the mean for each allele ranged from 20 to 27 Da, or approximately 0.1% relative standard deviation (RSD). The mass between alleles is equal to the repeat unit, which in the case of TPOX is 1,260 Da for an AATG repeat (exhibit 62). Thus, each allele is easily distinguishable.

Measurements were made of the same DNA samples over a fairly wide time-span, revealing that masses can be remarkably similar, even when data points are recollected months later. Exhibit 60 compares 57 allele meas-

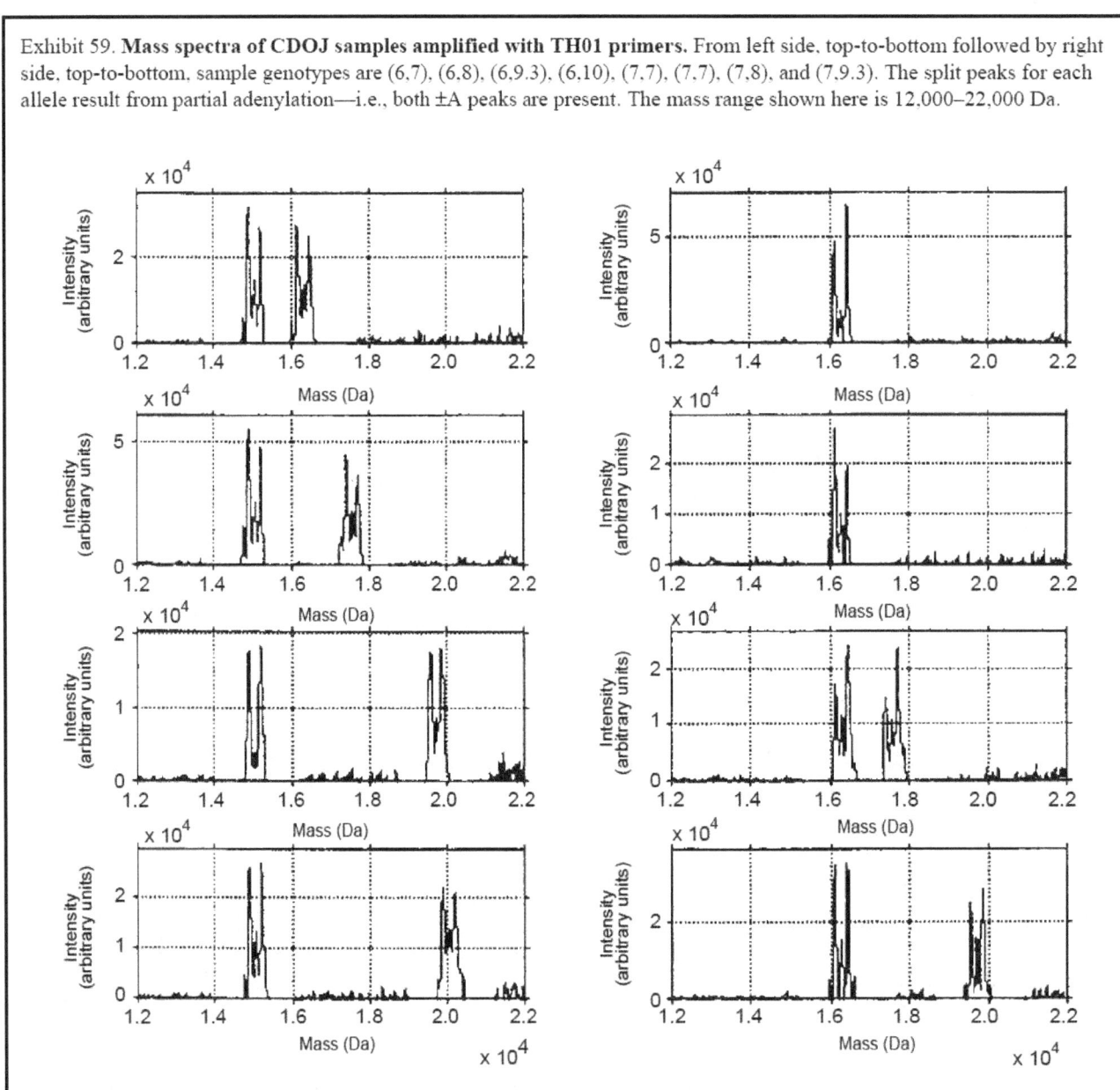

Exhibit 59. **Mass spectra of CDOJ samples amplified with TH01 primers.** From left side, top-to-bottom followed by right side, top-to-bottom, sample genotypes are (6,7), (6,8), (6,9.3), (6,10), (7,7), (7,7), (7,8), and (7,9.3). The split peaks for each allele result from partial adenylation—i.e., both ±A peaks are present. The mass range shown here is 12,000–22,000 Da.

Exhibit 60. **Comparison of allele masses collected 6 months apart.** This plot compares 57 allele measurements of 6 different TPOX alleles. The ideal line is shown on the same plot to demonstrate how reproducible the masses are over time. The average standard deviation of allele mass measurements between these 2 data sets was 47 Da. This result further confirms that no allelic ladders or other internal DNA standards are needed to obtain accurate measurements with mass spectrometry.

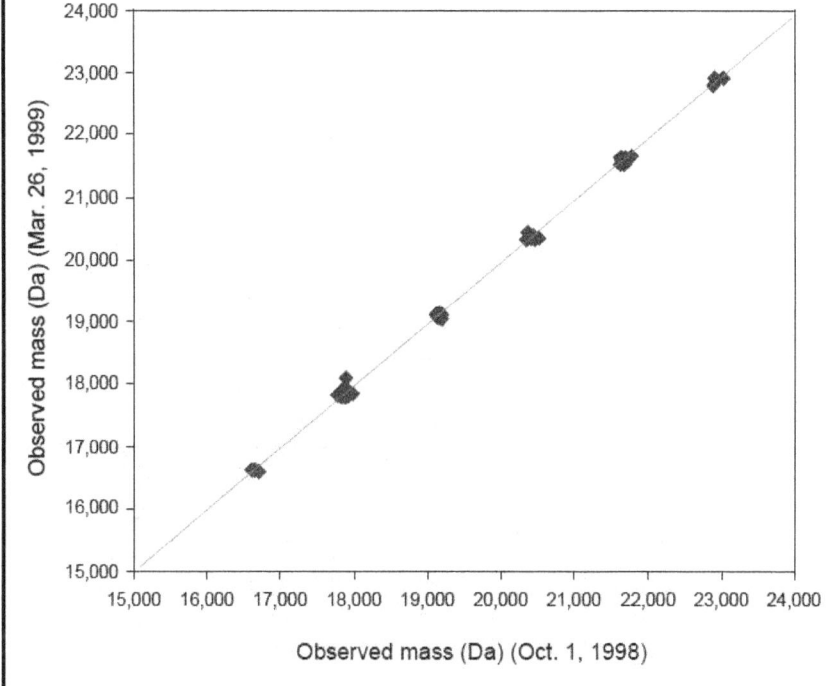

urements from 6 different TPOX alleles collected 6 months apart. The first data set was collected on October 1, 1998, and the second data set on March 26, 1999. Amazingly enough, some of the alleles had identical measured masses, even though different mass calibration constants (and even different instruments) were used.

The bottom line is whether or not a correct genotype can be obtained using this new technology. Exhibit 63 compares the genotypes obtained using a conventional CE separation method and this mass spectrometry technique across 3 STR markers (D16S539, D8S1179, and CSF1PO) and indicates an excellent agreement between the methods. With the CDOJ samples tested, there was complete agreement on all observed genotypes for the STR loci CSF1PO, TH01, and D3S1358 as well as the sex-typing marker amelogenin (exhibits 12, 14–16). Some "gas-phase" dimers and trimers fell into the allele mass range and confused the calling for TPOX (exhibit 13) and D16S539 (exhibit 17) on several samples. Gas-phase dimers and trimers are assay artifacts that result from multiple excess primer molecules colliding in the gas phase and being ionized during the MALDI process. A mass offset plot like that shown in exhibit 46 can be used to detect these assay artifacts as they fall outside the tight grouping and inside the 300 Da window. With the CDOJ samples, D7S820 exhibited null alleles (exhibit 18) and FGA had some unique challenges due to its larger size, such as problems with resolution of closely spaced heterozygotes and poorer mass calibration since the measured alleles were further away from the calibration standards (exhibit 19). Thus, when the PCR situations such as null alleles are accounted for and smaller loci are used, this mass spectrometry method produces results comparable to traditional methods of STR genotyping.

Data collection speed

The tremendous speed advantage of mass spectrometry can be seen in exhibit 11. Over the course of this project, data collection speed increased by a factor of 10 from ~50 seconds/sample to less than 5 seconds/sample. This speed increase resulted from improved software and hardware on the automated mass spectrometers and from improved sample quality (i.e., better PCR conditions that yielded more product and improved sample cleanup that in turn yielded "cleaner" DNA). With data collection time around 5 seconds per sample, achieving sample throughputs of almost 1,000 samples per hour is possible, and 3,000–4,000 samples per system per day is reasonable when operating at full capacity. Sample backlogs could be erased rather rapidly with this kind of throughput. By way of comparison, it takes an average of 5 minutes to obtain each genotype (assuming a multiplex level of 6 or 7 STRs) using conventional CE methods (exhibit 11). Thus, the mass spectrometry method described in this study is two orders of magnitude faster in sample processing time than conventional techniques.

59

Exhibit 61. **Fifteen replicate analyses of a TPOX allelic ladder to measure mass precision and accuracy.** The precision was less than 30 Da for a single standard deviation, which corresponds to less than 0.1 nucleotide. The measured mass accuracy from the calculated expected allele masses averaged ~30 Da. Across the 8 alleles in the ladder, 120 data points are used to make this determination. All numbers are in Daltons (Da). Percentage error was calculated as (observed-expected)/expected. This same data is also presented in histogram format; see figure 1 in Butler et al., 1998.

	Allele 6	Allele 7	Allele 8	Allele 9	Allele 10	Allele 11	Allele 12	Allele 13
Expected mass (Da)	15,345	16,605	17,865	19,125	20,385	21,644	22,904	24,164
Sample number								
1	15,346	16,623	17,903	19,130	20,388	21,623	22,860	24,074
2	15,387	16,667	17,901	19,129	20,387	21,639	22,893	24,114
3	15,372	16,629	17,887	19,143	20,400	21,615	22,877	24,091
4	15,385	16,653	17,903	19,163	20,384	21,642	22,903	24,076
5	15,388	16,642	17,898	19,155	20,337	21,654	22,870	24,111
6	15,336	16,600	17,857	19,105	20,362	21,599	22,832	24,064
7	15,388	16,637	17,894	19,131	20,383	21,635	22,904	24,110
8	15,363	16,618	17,872	19,129	20,368	21,604	22,853	24,087
9	15,365	16,628	17,891	19,150	20,385	21,620	22,892	24,087
10	15,373	16,638	17,892	19,136	20,394	21,631	22,878	24,085
11	15,383	16,640	17,896	19,152	20,387	21,621	22,884	24,129
12	15,388	16,648	17,912	19,172	20,388	21,623	22,850	24,149
13	15,407	16,660	17,941	19,208	20,425	21,674	22,944	24,148
14	15,410	16,659	17,930	19,174	20,425	21,666	22,893	24,132
15	15,390	16,648	17,915	19,157	20,423	21,636	22,897	24,126
Average mass	15,379	16,639	17,899	19,149	20,389	21,632	22,882	24,106
Std. dev.	20.2	17.8	20.6	24.7	23.7	21.0	27.2	27.3
%RSD	0.13	0.11	0.12	0.13	0.12	0.10	0.12	0.11
% error	0.22	0.21	0.19	0.13	0.02	-0.06	-0.10	-0.24
Obs-exp	33.7	34.3	34.5	23.9	4.1	-11.9	-22.0	-58.5

Exhibit 62. **Upper strand (TCAT repeat) and lower strand (AATG repeat) mass differences for the TH01 allelic ladder.** The upper strand was discernible from the lower strand due to the different sequence contents of the repeats. The STR repeat structure and nucleotide content can be seen using mass spectrometry. Note: The upper strand mass difference between alleles 9.3 and 10 is 306 Da, or a "T," and the lower strand mass difference between these same two alleles is 315 Da, or an "A." For more details, see Butler et al., 1998b.

Upper Strand	Expected (Da)	Observed (Da)
Allele 5–6	1,211	1,210
Allele 6–7	1,211	1,211
Allele 7–8	1,211	1,215
Allele 8–9	1,211	1,215
Allele 9–9.3	907	915
Allele 9.3–10	304	306
Allele 9–10	1,211	1,221

Repeat = TCAT = 1,210.8 Da
= --CAT = 906.6 Da

Upper Strand	Expected (Da)	Observed (Da)
Allele 5–6	1,260	1,259
Allele 6–7	1,260	1,262
Allele 7–8	1,260	1,269
Allele 8–9	1,260	1,267
Allele 9–9.3	947	948
Allele 9.3–10	313	315
Allele 9–10	1,260	1,263

Repeat = AATG = 1,259.8 Da
= --ATG = 946.6 Da

Exhibit 63. **Comparison of ABI 310 and mass spectrometry allele calls for 90 CEPH/diversity samples.** Out of 1,080 possible allele calls with these 3 STR loci, there were 100 with no data collected (indicated as a "0" on the allele call axis), and only 12 calls differed between the two methods, or ~98% correlation.

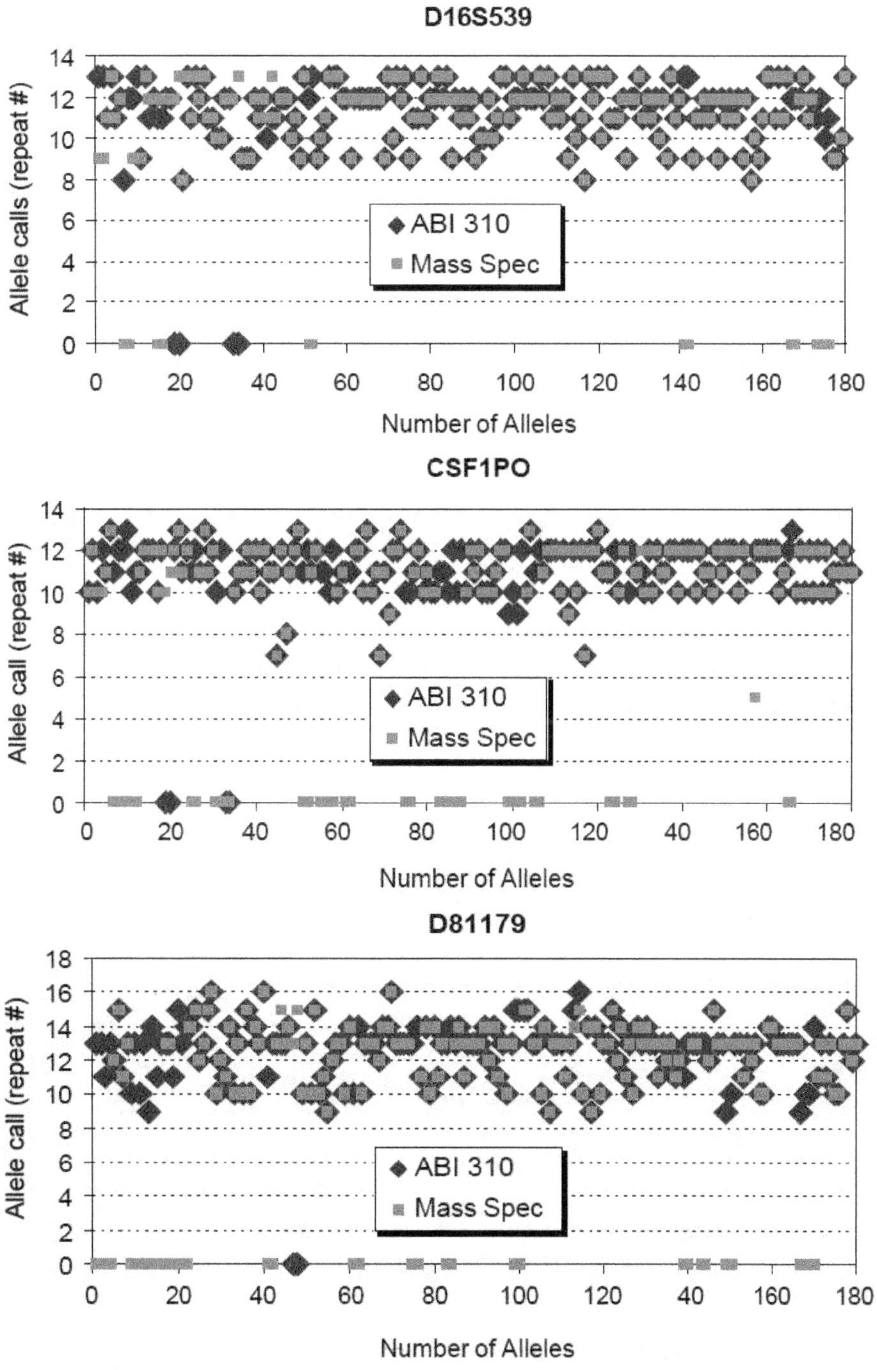

D16S539

CSF1PO

D81179

RESULTS AND DISCUSSION OF MULTIPLEX SNPs

Work began on the development of multiplexed SNP assays in the summer of 1998 after notice that a second NIJ grant, Development of Multiplexed Single Nucleotide Polymorphism Assays from Mitochondrial and Y-Chromosome DNA for Human Identity Testing Using Time-of-Flight Mass Spectrometry, had been funded. Excellent progress was made toward the milestones on this grant, but the work not finished because this grant was prematurely terminated on the part of GeneTrace in the spring of 1999. The completed work focused on two areas: the development of a 10-plex SNP assay from the mtDNA control region using a single amplicon and the development of a multiplex PCR assay from Y-chromosome SNP markers that involved as many as 18 loci amplified simultaneously. This section describes the design aspects of multiplex PCR and SNP assays along with the progress made toward the goal of producing assays that would be useful for high-throughput screening of mitochondrial and Y-chromosome SNP markers.

The approach to SNP determination described here has essentially three steps: (1) PCR amplification, (2) phosphatase digestion, and (3) SNP primer extension. Either strand of DNA may be probed simultaneously in this SNP primer extension assay. PCR primers are designed to generate an amplicon that includes one or more SNP sites. The initial PCR reaction is performed with standard (unlabeled) primers. A phosphatase is then added following PCR to remove all remaining dNTPs so that they will not interfere with the single base extension reaction involving ddNTPs. These reactions can all be performed in the same tube or well in a sample tray. A portion of the phosphatase-treated PCR product is then used for the primer extension assay.

In the SNP primer extension assay, a special primer containing a biotin moiety at the 5'-end permits solid-phase capture for sample purification prior to mass spectrometry analysis. This primer hybridizes upstream of the SNP site with the 3'-end immediately adjacent to the SNP polymorphic site. A cleavable nucleotide near the 3'-end allows the 3'-end of the primer to be released from the immobilized portion and reduces the overall mass of the measured DNA molecule (exhibit 21) (Li et al., 1999). The complementary nucleotide(s) to the nucleotide(s) present at the SNP site is inserted during the extension reaction. In the case of a heterozygote, two extension products result. Only a single base is added to the primer during this process because only ddNTPs are used and the dNTPs left over from PCR are hydrolyzed with the phosphatase digestion step. If the extension reaction is not driven to completion (where the primer would be totally consumed), then both primer and extension product (i.e., primer plus single nucleotide) are present after the primer extension reaction. The mass difference between these two DNA oligomers is used to determine the nucleotide present at the SNP site. In the primer extension SNP assay, the primer acts as an internal standard and helps make the measurement more precise. A histogram of mass difference measurements across 200 samples (50 per nucleotide) is shown in exhibit 64. The ddT and ddA differ by only 9 Da and are the most difficult to resolve as heterozygotes or distinguish from one

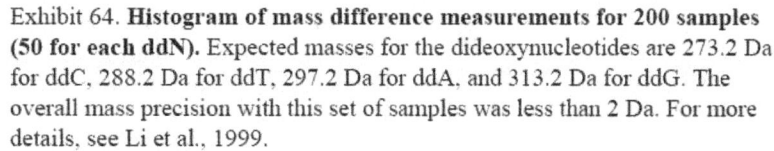

Exhibit 64. **Histogram of mass difference measurements for 200 samples (50 for each ddN).** Expected masses for the dideoxynucleotides are 273.2 Da for ddC, 288.2 Da for ddT, 297.2 Da for ddA, and 313.2 Da for ddG. The overall mass precision with this set of samples was less than 2 Da. For more details, see Li et al., 1999.

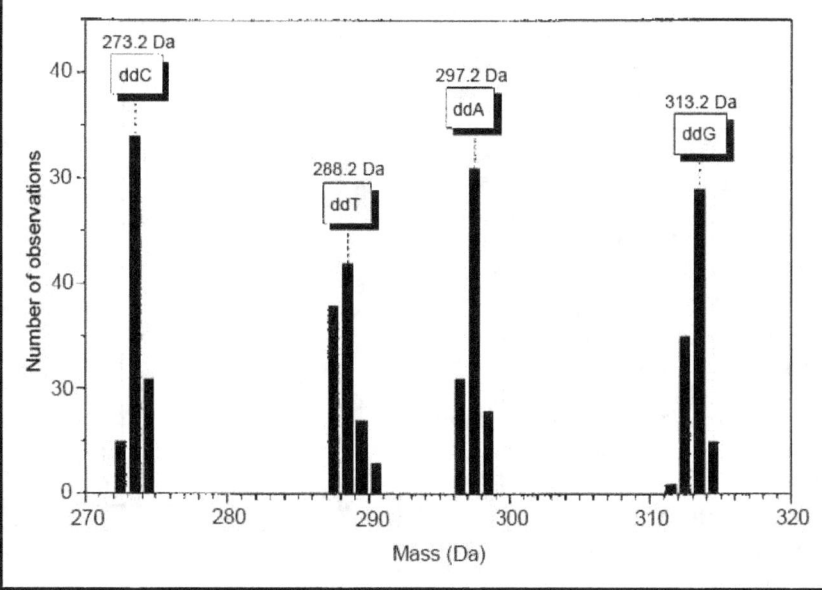

another in terms of mass. As reported in a recently published paper (Li et al., 1999), this approach has been used to reliably determine all four possible SNP homozygotes and all six possible heterozygotes.

Mitochondrial DNA Work

The control region of mtDNA, commonly referred to as the D-loop, is highly polymorphic and contains a number of possible SNP sites for analysis. MITOMAP, an internet database containing fairly comprehensive information on mtDNA, lists 408 polymorphisms over 1,121 nucleotides of the control region (positions 16020–576) that have been reported in literature (MITOMAP, 1999). However, many of these polymorphisms are rare and population specific. The present study focused on marker sets from a few dozen well-studied potential SNP sites. Special Agent Mark Wilson from the FBI Laboratory in Washington, D.C., who has been analyzing mtDNA for more than 7 years, recommended a set of 27 SNPs that would give a reasonable degree of discrimination and make the assay about half as informative as full sequencing. His recommended mtDNA sites were positions 16069, 16114, 16126, 16129, 16189, 16223, 16224, 16278, 16290, 16294, 16296, 16304, 16309, 16311, 16319, 16362, 73, 146, 150, 152, 182, 185, 189, 195, 198, 247, and 309. The underlined sites are those reported in a minisequencing assay developed by the Forensic Science Service (Tully et al., 1996). GeneTrace's multiplex SNP typing efforts began with 10 of the SNPs used in the FSS minisequencing assay, since those primer sequences had already been reported and studied together.

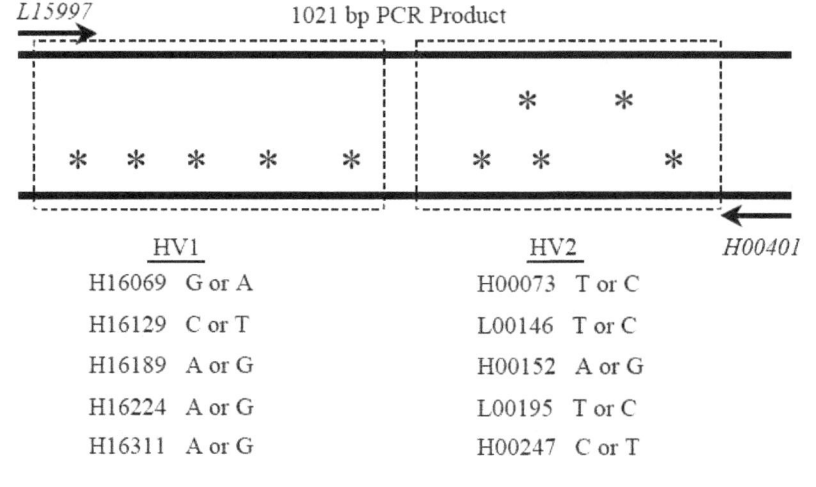

Exhibit 65. **Schematic representation of the mtDNA control region 10-plex SNP assay.** The asterisks represent the relative positions of the SNP sites and the strand that is probed.

HV1		HV2	
H16069	G or A	H00073	T or C
H16129	C or T	L00146	T or C
H16189	A or G	H00152	A or G
H16224	A or G	L00195	T or C
H16311	A or G	H00247	C or T

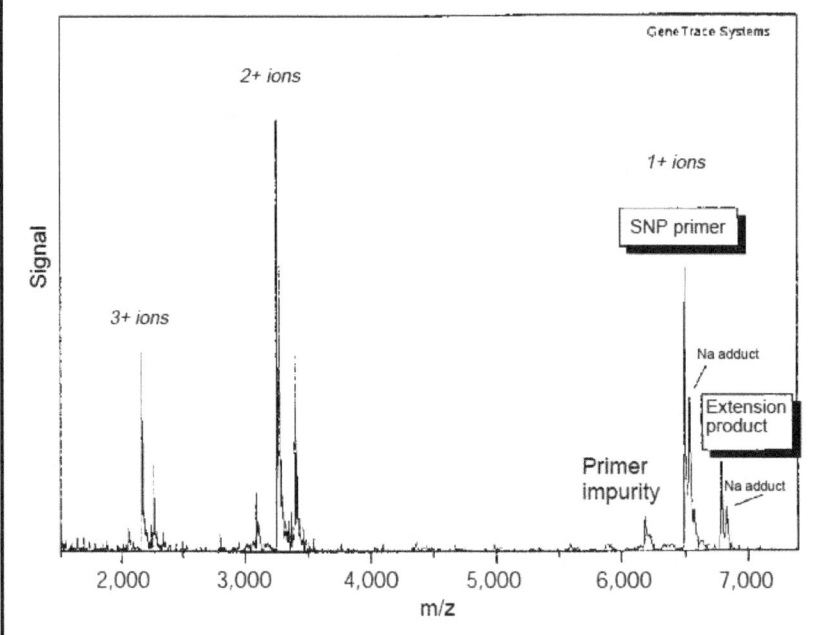

Exhibit 66. **SNP ions impacting multiplex design.** Doubly charged and triply charged ions from higher mass primers can interfere with singly charged ions of smaller mass primers if the multiplex is not well designed.

The reported FSS sequences were modified slightly by removing the poly(T) tail and converting the degenerate bases into the most common sequence variant (identified by examination of MITOMAP information at the appropriate mtDNA position). A cleavable base was also incorporated at varying positions in different primers so that the cleaved primers would be resolvable on a mass scale. Exhibit 26 lists the final primer set chosen for a 10-plex SNP reaction. Eight of the primers detected SNPs on the "heavy" GC-rich strand and two of them identified SNPs on the "light" AT-rich strand of the mtDNA control region (exhibit 65). Five of the primers annealed within hypervariable region I (HV1) and five annealed within hypervariable region II (HV2). All of the 10 chosen SNP sites were transitions of either A to G (purine-to-purine) or C to T (pyrimidine-to-pyrimidine) rather than transversions (purine-to-pyrimidine).

Besides primer compatibility (i.e., lack of primer dimer formation or hairpins), another important aspect of multiplex SNP primer design is the avoidance of multiple-charged ions. Doubly charged and triply charged ions of larger mass primers can fall within the mass-to-charge range of smaller primers. Depending on the laser energy used and matrix crystallization, the multiple-charged ions can be significantly abundant (exhibit 66). Primer impurities, such as n-1 failure products, can also impact how close together primers can be squeezed on a mass scale. These primer synthesis failure products will be ~300 Da smaller in mass than the full-length primer. Since an extension product ranges from 273 (ddC) to 313 Da (ddG) larger than the primer itself, a minimum of 650–700 Da is needed between adjacent primers (postcleavage mass) if primer synthesis failures exist to

avoid any confusion in making the correct SNP genotype call.

Primer synthesis failure products were observed to become more prevalent for larger mass primers. Because resolution and sensitivity in the mass spectrometer decrease at higher masses, it is advantageous to keep the multiplexed primers in a fairly narrow mass window and as small as possible. The primers in this study ranged from 1,580 to 6,500 Da. Exhibit 23 displays the expected primer masses for the mtDNA SNP 10-plex along with their doubly and triply charged ions. The smallest four primers, in the mass range of 1,580–3,179 Da, had primer and extension masses that were similar to multiple-charged ions of larger primers. For example, in the bottom panel of exhibit 6, which shows the 10-plex primers, the doubly charged ion from MT4e (3,250 Da), which probes site L00195, fell very close to the singly charged ion from MT3' (3,179 Da), which probes site H16189. The impact of primer impurity products can also be seen in exhibit 6. An examination of the extension product region from primer MT7/H00073 (~6,200 Da) shows two peaks where only one was expected (top panel). The lower mass peak in the doublet is labeled as "+ddC" (6,192 Da), but the larger peak in the doublet was a primer impurity of MT4e/L00195 (6,232 Da). The mass difference between these two peaks was 40 Da or exactly what one would expect for a C/G heterozygote extension of primer MT7/H00073. Thus, to avoid a false positive, it was important to run the 10 primers alone as a negative control to verify any primer impurities.

To aid development of this multiplex SNP assay, large quantities of PCR product were produced from K562 genomic DNA (enough for ~320 reactions) and were pooled together so that multiple experiments would have

the same starting material. With the K562 amplicon pool, the impact of primer concentration variation was examined without worrying about the DNA template as a variable. The K562 amplicon pool was generated using the PCR primers noted in exhibit 26, which produced a 1,021 bp PCR product that spanned the entire D-loop region (Wilson et al., 1995). Thus, all 10 SNP sites could be examined from a single DNA template.

Using ABI's standard sequencing procedures and dRhodamine dye-terminator sequencing kit, this PCR pool was sequenced to verify the identity of the nucleotide at each SNP site in the 10-plex. The sequencing primers were the same as those reported previously (Wilson et al., 1995). Identical results were obtained between the sequencing and the mass spectrometer, which verified the method (exhibit 6).

A variety of primer combinations and primer concentrations were tested on the way to obtaining results with the 10-plex. For example, a 4-plex and a 6-plex were developed first with primers that were further apart in terms of mass and, therefore, could be more easily distinguished. An early 6-plex was published in *Electrophoresis* (Li et al., 1999). Primer concentrations were balanced empirically by first running all primers at 10 pmol and then raising or lowering the amount of primer in the next set to obtain a good balance between those in the multiplex primer mix. In general, a higher amount of primer was required for primers of higher mass. However, this trend did not always hold true, probably because ionization efficiencies in MALDI mass spectrometry differ depending on DNA sequence content. The primer concentrations in the final "optimized" 10-plex ranged from 10 pmol for MT3' (3,179 Da) to 35 pmol for MT7 (5,891 Da). Primer extension efficiencies also varied between primers, making optimization of these multiplexes rather challenging.

Originally, this study set out to examine 100 samples, but due to the early termination of the project, researchers were unable to run this multiplex SNP assay across a panel of samples to verify that it worked with more than one sample. Future work could include examination of a set of population samples and correlation to DNA sequencing results. Examination of the impact of SNPs that are close to the one being tested and that might impact primer annealing also needs to be done. In addition, more SNP sites can be developed and the multiplex could be expanded to include a larger number of loci.

Y-Chromosome Work

While the mtDNA work produced an opportunity to examine the mass spectrometry factors in developing an SNP multiplex, this work involved only a single DNA template with multiple SNP probes. A more common situation for multiplex SNP development is multiple DNA templates with one or more SNP per template. SNP sites may not be closely spaced along the genome and could require unique primer pairs to amplify each section of DNA. To test this multiplex SNP situation, researchers investigated multiple SNPs scattered across the Y chromosome. Through a collaboration with Dr. Oefner and Dr. Underhill, 20 Y-chromosome SNP markers were examined in this study. Dr. Oefner and Dr. Underhill have identified almost 150 SNP loci on the Y chromosome, some of which have been reported in the literature (Underhill et al., 1997). By examining an initial set of 20 Y SNPs and adding additional markers as needed, researchers attempted to develop a final multiplex set based on ~50 Y SNP loci. The collaboration provided

detailed sequence information on bases around the SNP sites (typically several hundred bases on either side of the SNP site), which is important for multiplex PCR primer design. Dr. Oefner also provided a set of 38 male genomic DNA samples from various populations around the world for testing purposes.

The sequences were provided in two batches of 10 sequences each. In the first two sets, primer designs were attempted for a 9-plex PCR and a 17-plex PCR, respectively. Due to primer incompatibilities, it was impossible to

incorporate all SNPs into each multiplex set. However, with a larger set of sequences to choose from, it is conceivable that much larger PCR multiplexes can be developed. According to Dr. Underhill's nomenclature, the first set of Y SNP markers included the following loci: M9 (C→G), M17 (1 bp deletion, 4Gs→3Gs), M35 (G→C), M42 (A→T), M45 (G->A), M89 (C→T), M96 (G→C), M122 (T→C), M130 (C→T), and M145 (G→A). The second set of Y SNP markers contained these loci: M119 (A→C), M60 (1 bp insertion, a "T"), M55 (T→C),

---

Exhibit 67. **Multiplex PCR information for 17-plex PCR reaction**

Locus Name	Expected SNP	Primers Used	Primer Amounts	PCR Product Size (bp)
M3	C/T	F1u/R1u	0.4 pmole each	148 bp
M17	1 bp del	F1u/R1u	0.4 pmole each	149 bp
M13	G/C	F1u/R1u	0.4 pmole each	172 bp
M119	A/C	F1u/R1u	0.4 pmole each	188 bp
M2	A/G	F3u/R3u	0.4 pmole each	194 bp
M96	G/C	F1u/R1u	0.4 pmole each	198 bp
M122	T/C	F2u/R2u	0.4 pmole each	208 bp
M145	G/A	F1u/R1u	0.4 pmole each	218 bp
M45	G/A	F2u/R2u	0.4 pmole each	230 bp
M35	G/C	F1u/R5u	0.4 pmole each	233 bp
M9	C/G	F1u/R1u	0.4 pmole each	243 bp
M55	T/C	F1u/R1u	0.4 pmole each	247 bp
M60	1 bp ins	F1u/R1u	0.4 pmole each	263 bp
M89	C/T	F4u/R4u	0.4 pmole each	275 bp
M42	A/T	F1u/R1u	0.4 pmole each	300 bp
M67	A/T	F1u/R1u	0.4 pmole each	314 bp
M69	T/C	F1u/R1u	0.4 pmole each	326 bp
M26	G/C	F1u/R1u	0.4 pmole each	333 bp
M130	C/T	F1u/R1u	Included in 9-plex but not 17-/18-plex PCR	155 bp

M20 primers are not compatible with the other loci in the multiplex.

PCR Mix: 5 mM MgCl$_2$, 1X PCR buffer II, 250 µM dNTPs, 2 U TaqGold, 40 pmol univ-F primer, 40 pmol univ-R primer, and 0.4 pmole each locus-specific primer pair in 20 µL volume

Thermal Cycling: 95 °C for 10 min; 50 cycles at 94 °C for 30 sec, 55 °C for 30 sec, 68 °C for 60 sec, 72 °C for 5 min, and 4 °C forever (~3 hr total)

M20 (A→G), M69 (T→C), M67 (A→T), M3 (C→T), M13 (G→C), M2 (A→G), and M26 (G→A).

Multiplex PCR primers were designed with a UNIX version of Primer 3 (release 0.6) (Rozen et al., 1998) that was adapted at GeneTrace by Nathan Hunt to utilize a mispriming library and Perl scripts for input and export of SNP sequences and primer information, respectively. The PCR primer sequences produced by Hunt's program are listed in exhibit 27. The universal tags attached to each primer sequence aid in multiplex compatibility (Shuber et al., 1995). This tag added 23 bases to the 5'-end of the forward primers and 24 bases to the 5'-end of the reverse primers and, therefore, increased the overall length of PCR products by 47 bp. The addition of the universal tag makes multiplex PCR development much easier and reduces the need to empirically adjust primer concentrations to balance PCR product quantities obtained from multiple loci (Ross et al., 1998).

To compare the amplicon yields from various loci amplified in the multiplex PCR, the product sizes were selected to make them resolvable by CE separation. Thus, the PCR product sizes ranged from 148 bp to 333 bp (exhibit 67) using the primers listed in exhibit 27. To make sure that each primer pair worked, each marker was amplified individually as well as in the multiplex set using the same concentration of PCR primers. A substantial amount of primers remaining after PCR indicated that the PCR efficiency was lower for that particular marker (exhibit 68). Researchers were able to demonstrate male-specific PCR with a 17-plex set of PCR primers. The male test sample AM209 from an Amish CEPH family (exhibit 25) produced amplicons for 17 Y SNP loci, while K562 genomic DNA yielded no detectable PCR prod-

uct because it is female DNA and, therefore, does not contain a Y chromosome (exhibit 7).

SNP primers were designed and synthesized for probing the SNP sites either in a singleplex (exhibit 69) or a multiplex (exhibit 70) format. Some additional SNP primers and multiplex PCR primers were also designed for testing 12 autosomal SNPs throughout the human genome with the hope of comparing the informativeness of SNPs to STRs (exhibit 71). Analysis of these same 12 SNPs was recently demonstrated in a multiplex PCR and SNP assay by PerSeptive Biosystems (Ross et al., 1998).

Optimal SNP markers for identity testing typically have allele frequencies of 30–70% in a particular human population. By way of comparison, highly polymorphic STRs can have 10–15 or more alleles with allele frequencies below 15% (i.e., more alleles and lower allele frequencies). The characteristics of STR and SNP markers are compared in exhibit 72. SNPs have the capability of being multiplexed to a much higher level than STRs; however, more SNP markers are required for the same level of discrimination compared with STRs. Only time will tell what role new SNP markers will have in human identity testing.

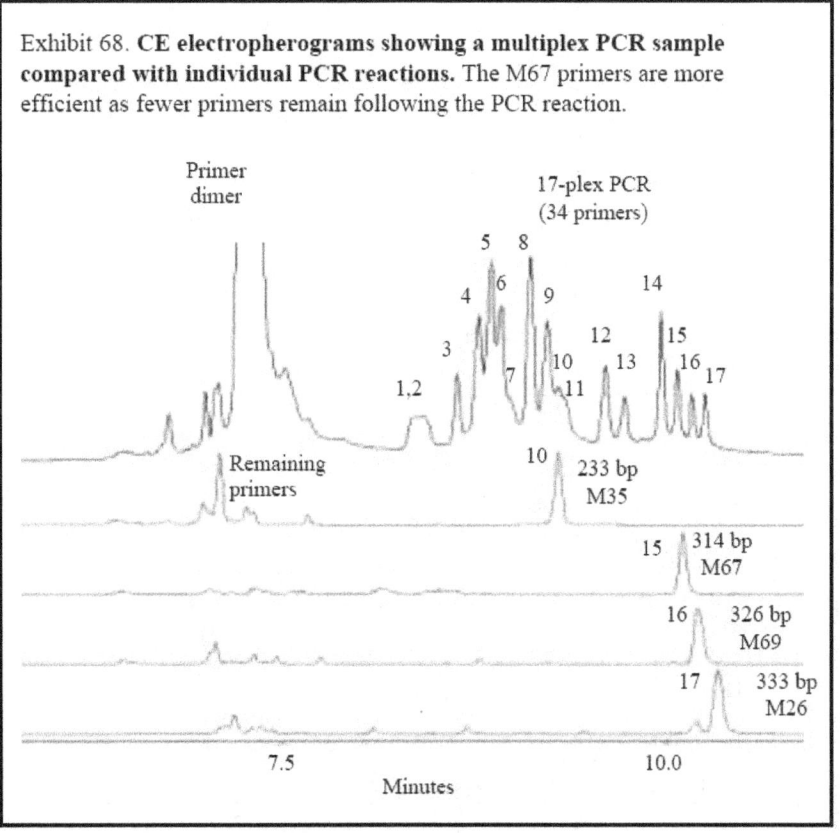

Exhibit 68. **CE electropherograms showing a multiplex PCR sample compared with individual PCR reactions.** The M67 primers are more efficient as fewer primers remain following the PCR reaction.

Exhibit 69. **Primers for testing Y-chromosome SNP markers in singleplexes.** Primers were designed automatically with UNIX SNP probe scripts written by Nathan Hunt but have not yet been tested.

Primer Name	Primer Sequence (5'→3')	Cleavage Mass (Da)	Expected SNP
M145-P1	b-CTTGCCTCCACGAC(T)TTCCT	1,491	A/G
M35-P1	b-CGGAGTCTCTGCC(T)GTGTC	1,556	C/G
M9-P1	b-AACGGCCTAAGATGG(T)TGAAT	1,564	C/G
M26-P1	b-AGGCCATTCAGTG(T)TCTCTG	1,820	C/G
M67-P1	b-TTGTTCGTGGACCCC(T)CTATAT (overlaps PCR reverse primer)	1,828	A/T
M45-P1	b-CCTCAGAAGGAGC(T)TTTTGC	1,835	C/T
M145-P2	b-GATTAGGCTAAGGC(T)GGCTCT	1,845	C/T
M119-P1	b-TTCCAATTCAGCA(T)ACAGGC	1,863	T/G
M55-P1	b-GCCCCTGGATGGTT(T)AAGTTA	1,877	C/T
M20-P1	b-ACCAACTGTGGAT(T)GAAAAT (no PCR primers designed)	1,886	A/G
M122-P1	b-TCAGATTTTCCCC(T)GAGAGC	1,903	A/G
M96-P1	b-TTGGAAAACAGGTCTC(T)CATAATA	2,150	C/G
M69-P1	b-GAGGCTGTTTACAC(T)CCTGAAA	2,151	C/T
M130-P1	b-GGGCAATAAACCT(T)GGATTTC (overlaps PCR forward primer)	2,173	C/T
M42-P1	b-CACCAGCTCTCTTTTTCAT(T)ATGTAGT	2,198	A/T
M89-P1	b-CAACTCAGGCAAAG(T)GAGAGAT (overlaps PCR reverse primer)	2,232	A/G
M3-P1	b-GGGTCACCTC(T)GGGACTGA	2,537	A/G
M2-P1	b-CCTTTATCC(T)CCACAGATCTCA (overlaps PCR reverse primer)	3,636	C/T
M13	Not designed		C/G
M60	Not designed (1 bp insertion)		ins
M17	Not designed (1 bp deletion)		del

Exhibit 70. **Y SNP multiplex primer information.** Primers were designed manually for multiplex SNP assay (9-plex) with nonoverlapping masses.

Primer Name	Primer Sequence (5'→3')	Cleaved Mass (Da)	Expected SNP
Y1 (M42b-a)	b-CCAGCTCTCTTTTTCATTA(T)GTAGT	1,580	T/A
Y2 (M96t-b)	b-CTTGGAAAACAGGTCTC(T)CATAATA	2,150	C/G
Y3 (M35b-c)	b-TTCGGAGTCTC(T)GCCTGTGTC	2,768	C/G
Y4 (M130b-b)	b-CCTT(T)CCCCTGGGCAG	3,380	C/T
Y5 (M145b-b)	b-GATTAGGC(T)AAGGCTGGCTCT	3,723	A/G
Y6 (M122t-c)	b-TAGAAAAGCAAT(T)GAGATACTAATTCA	4,333	C/T
Y7 (M45t-d)	b-AAATTGGCAG(T)GAAAAATTATAGATA	4,694	A/G
Y8 (M9b-c)	b-ACATGTCTAAA(T)TAAAGAAAAATAAAGAG	5,354	C/G
Y9 (M89t-e)	b-CTTCC(T)AAGGTTATGTACAAAAATCT	6,210	C/T

Exhibit 71. **Human autosomal SNP markers designed for testing NIH diversity panel**

SNP markers are not designed for multiplexing.

SNP Primer Name	Primer Sequence (5'→3')	Cleavage Mass (Da)	Expected (SNP)
C6-P1	b-GGGGACAGCCA(T)GCACTG	1,854	A/C
A2M-P1	b-GAAACACAGCAGCTTAC(T)CCAGAG	1,863	A/G
LDLR-P1	b-CCTATGACACCGTCA(T)CAGCAG	1,863	A/G
IL1A-P4	b-TTTTAGAAATCATCAAGCC(T)AGGTCA	1,878	T/G
CD18-P2	b-GGACATAGTGACCG(T)GCAGGT	1,894	C/T
IGF2-P1	b-CCACCTGTGATT(T)CTGGGG	1,910	C/T
ALDOB-P1	b-CGGGCCAAGAAGG(T)ATCTACC	2,102	A/G
PROS1-P1	b-CATAATGATATTAGAGCTCAC(T)CATGTCC	2,118	A/G
NF1-P1	b-CGATGGTTGTATTTGTCACCA(T)ATTAATT	2156	A/G
AT3-P0	b-ATCTCCA(T)GGGCCCAGC	2,787	C/T
CYP2D6-P1	b-GCAGCTTCAATGA(T)GAGAACCTG	2,810	C/T
LIPC-P0	b-AACATGGCT(T)CGAGAGAGTTG	3,483	A/C

Multiplex PCR primers for 12-plex PCR (not all amplicons are resolvable by CE). Universal primer sequences are in color.

Primer Name	Primer Sequence (5'→3')	PCR Product Size
AT3-F1u	ATT TAG GTG ACA CTA TAG AAT ACT GAG ACC TCA GTT TCC TCT TCT G	159 bp
AT3-R1u	TAA TAC GAC TCA CTA TAG GGA GAC CCT GGT CCC ATC TCC TCT AC	
C6-F1u	ATT TAG GTG ACA CTA TAG AAT ACA TCT GTC TTG CGT CCC AGT C	160 bp
C6-R1u	TAA TAC GAC TCA CTA TAG GGA GAC TCT TGC AGT CAG CCT CTT CA	
IGF2-F1u	ATT TAG GTG ACA CTA TAG AAT ACA GTC CCT GAA CCA GCA AAG A	163 bp
IGF2-R1u	TAA TAC GAC TCA CTA TAG GGA GAC TTT TCG GAT GGC CAG TTT AC	
LIPC-F1u	ATT TAG GTG ACA CTA TAG AAT ACA ACA CAC TGG ACC GCA AAA G	173 bp
LIPC-R1u	TAA TAC GAC TCA CTA TAG GGA GAC ACC CAG GCT GTA CCC AAT TA	
NF1-F1u	ATT TAG GTG ACA CTA TAG AAT ACA AGG AGC AAA CGA TGG TTG TA	181 bp
NF1-R1u	TAA TAC GAC TCA CTA TAG GGA GAC TAG GTG GCT GCA AGG TAT CC	
LDLR-F5u	ATT TAG GTG ACA CTA TAG AAT ACC CAC GGC GTC TCT TCC TAT	181 bp
LDLR-R5u	TAA TAC GAC TCA CTA TAG GGA GAC TGG TAT CCG CAA CAG AGA CA	
CYP2D6-F1u	ATT TAG GTG ACA CTA TAG AAT ACG GTG CAG AAT TGG AGG TCA T	182 bp
CYP2D6-R1u	TAA TAC GAC TCA CTA TAG GGA GAC AGA ACA GGT CAG CCA CCA CTA	
CD18-F2u	ATT TAG GTG ACA CTA TAG AAT ACA TCC AGG AGC AGT CGT TTG T	193 bp
CD18-R2u	TAA TAC GAC TCA CTA TAG GGA GAC ATG CCG CAC TCC AAG AAG	
ALDOB-F1u	ATT TAG GTG ACA CTA TAG AAT ACC ACA TTT GGG GCT TGA CTT T	231 bp
ALDOB-R1u	TAA TAC GAC TCA CTA TAG GGA GAC TCC TTC AGT CTC CTG TCA TCA A	
A2M-F1u	ATT TAG GTG ACA CTA TAG AAT ACC TCT GCC ATG CAA AAC ACA C	247 bp
A2M-R1u	TAA TAC GAC TCA CTA TAG GGA GAC AAC ATT CAA GTT TCC CTT ACT CAA	
PROS1-F1u	ATT TAG GTG ACA CTA TAG AAT ACT AAT GGC TGC ATG GAA GTG A	292 bp
PROS1-R1u	TAA TAC GAC TCA CTA TAG GGA GAC CCA GGA AAG GAC CAC AAA AT	
IL1A-F1u	ATT TAG GTG ACA CTA TAG AAT ACT TTG CTT CCT CAT CTG GAT TG	324 bp
IL1A-R1u	TAA TAC GAC TCA CTA TAG GGA GAC AGC AGC CGT GAG GTA CTG AT	

Exhibit 72. **Characteristics of STR and SNP markers.** SNPs are more common in the human genome than STRs but are not as polymorphic.

Characteristics	Short Tandem Repeats (STRs)	Single Nucleotide Polymorphisisms (SNPs)
Occurrence in human genome	~1 in every 15 kb	~1 in every 1 kb
General informativeness	High	Low (20–30% as informative as STRs)
Marker type	Di-, tri-, tetranucleotide repeat markers	Biallelic markers
Number of alleles per marker	Typically >5	Typically 2
Current detection methods	Gel/capillary electrophoresis	Microchip hybridization
Multiplex capability (fluorescence)	>10 markers with multiple spectral channels	Potentially thousands on microchip
Heterozygote resolution (fluorescence)	Mobility differences between alleles	Spectral differences between labeled nucleotides
Mass spectrometry measurement	Mass measurement of PCR-amplified allele(s)	Mass difference between primer and extension product

Exhibit 73. **STR genotypes for standard DNA templates K562, AM209, and UP006 obtained using AmpF1STR® ProfilerPlus™ and AmpF1STR® COfiler™ fluorescent STR kits.** These samples were the primary controls used for testing of newly developed primer sets. Alleles were not always well balanced for the K562 cell line DNA due to possible chromosome imbalances in the original sample. For example, the STR marker D21S11 produced three balanced alleles, most likely because three chromosome 21s were present in the K562 genomic DNA. In addition, the allele 24 peak ("24w" = "weak") for the FGA locus in K562 is only one-third the height of the allele 21 peak probably because there are more copies of the chromosome 4 on which the allele with 21 repeats resides. K562 genomic DNA came from Promega Corporation and AM209 came from CEPH pedigree 884 (Amish family). UP006 is an anonymous population sample of European origin purchased from Bios Laboratory.

STR locus	K562	AM209	UP006
TH01	9.3,9.3	6,7	7,9.3
TPOX	8,9	8,9	8,8
CSF1PO	9,10	10,12	11,14
D8S1179	12,12	11,13	13,13
D21S11	29,30,31	30,30	28,31
D18S51	15,16	12,12	13,13
D3S1358	16,16	15,18	15,15
VWA	16,16	18,18	16,17
FGA	21,24w	20,25	19,22
D16S539	11,12	11,12	12,13
D5S818	11,12	11,12	10,11
D13S317	8,8	14,14	11,14
D7S820	9,11	9,10	10,12
Amelogenin	X,X	X,Y	X,Y

Exhibit 74. **An ABI 310 CE electropherogram showing a heterozygous sample with a 15.2 microvariant at STR locus D3S1358.** A high degree of resolution is often necessary to resolve two closely spaced alleles in which one of them is a microvariant. In this case, the two alleles differ by only 2 bp even though the core repeat is a tetranucleotide. In the comparison with STR data collected by CDOJ, this was the only sample where a different result was obtained between our analysis and that performed by CDOJ on the same STR markers. CDOJ reported 14,15 for this sample.

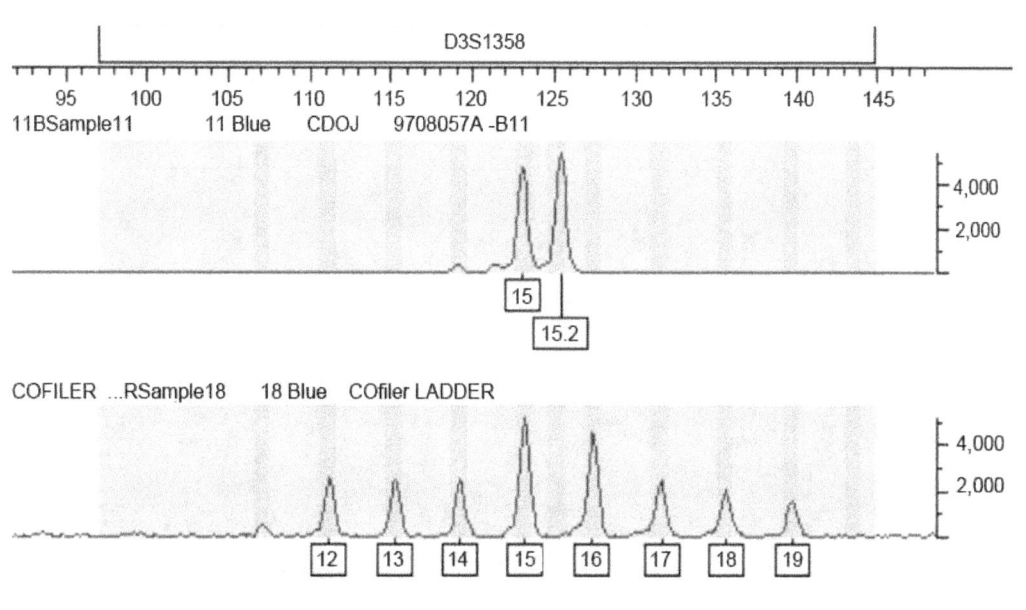

REFERENCES

ABI Prism 310 Genetic Analyzer User's Manual. 1998. Foster City, CA: Applied Biosystems.

AmpF1STR® COfiler™ PCR Amplification Kit User Bulletin. 1998. Foster City, CA: Applied Biosystems.

AmpF1STR® Profiler Plus™ PCR Amplification Kit User's Manual. 1998. Foster City, CA: Applied Biosystems.

Becker, C.H., J. Li, T.A. Shaler, J.M. Hunter, H. Lin, and J.A. Monforte. 1997. Genetic analysis of short tandem repeat loci by time-of-flight mass spectrometry. In *Proceedings from the Seventh International Symposium on Human Identification 1996.* Madison, WI: Promega Corporation, 158–162.

Becker, C.H. and S.E. Young. 1999. Mass spectrometer. U.S. Patent No. 5,864,137.

Braun, A., D.P. Little, and H. Koster. 1997a. Detecting CFTR gene mutations by using primer oligo base extension and mass spectrometry. *Clinical Chemistry* 43:1151–1158.

Braun, A., D.P. Little, D. Reuter, B. Muller-Mysok, and H. Koster. 1997b. Improved analysis of microsatellites using mass spectrometry. *Genomics* 46:18–23.

Butler, J.M., J. Li, T.A. Shaler, J.A. Monforte, and C.H. Becker. 1998. Reliable genotyping of short tandem repeat loci without an allelic ladder using time-of-flight mass spectrometry. *International Journal of Legal Medicine* 112(1):45–49.

Butler, J.M., B.R. McCord, J.M. Jung, J.A. Lee, B. Budowle, and R.O. Allen. 1995. Application of dual internal standards for precise sizing of polymerase chain reaction products using capillary electrophoresis. *Electrophoresis* 16:974–980.

Butler, J.M., J. Li, J.A. Monforte, and C.H. Becker. 2000. DNA typing by mass spectrometry with polymorphic DNA repeat markers. U.S. Patent No. 6,090,558.

Carroll, J.A. and R.C. Beavis. 1996. Using matrix convolution filters to extract information from time-of-flight mass spectra. *Rapid Communications in Mass Spectrometry* 10:1683–1687.

Clark, J.M. 1988. Novel non-templated nucleotide addition reactions catalysed by procaryotic and eucaryotic DNA polymerases. *Nucleic Acids Research,* 16(20):9677–9686.

Collins, F.S., L.D. Brooks, and A. Chakravarti. 1998. A DNA polymorphism discovery resource for research on human genetic variation. *Genome Research,* 8(12):1229–1231.

Fregeau, C.J. and R.M. Fourney. 1993. DNA typing with fluorescently tagged short tandem repeats: A sensitive and accurate approach to human identification. *BioTechniques* 15(1):100–119.

GenePrint™ STR Systems Technical Manual. 1995. Madison, WI: Promega Corporation, Part# TMD004.

Haff, L.A. and I.P Smirnov. 1997. Single-nucleotide polymorphism identification assays using a thermostable DNA polymerase and delayed extraction MALDI-TOF mass spectrometry. *Genome Research* 7(4):378–388.

Hammond, H.A., L. Jin, Y. Zhong, C.T. Caskey, and R. Chakraborty. 1994. Evaluation of 13 short tandem repeat loci for use in personal identification applications. *American Journal of Human Genetics* 55:175–189.

Kimpton, C.P., P. Gill, A. Walton, A. Urquhart, E.S. Millican, and M. Adams. 1993. Automated DNA profiling employing multiplex amplification of short tandem repeat loci. *PCR Methods and Applications* 3:13–22.

Lazaruk, K., P.S. Walsh, F. Oaks, D. Gilbert, B.B. Rosenblum, S. Menchen, D. Scheibler, H.M. Wenz, C. Holt, and J. Wallin. 1998. Genotyping of forensic short tandem repeat (STR) systems based on sizing precision in a capillary electrophoresis instrument. *Electrophoresis* 19(1):86–93.

Li, H., L. Schmidt, M.-H. Wei, T. Hustad, M.I. Lerman, B. Zbar, and K. Tory. 1993. Three tetranucleotide polymorphisms for loci: D3S1352, D3S1358, D3S1359. *Human Molecular Genetics* 2:1327.

Li, J., J.M. Butler, Y. Tan, H. Lin, S. Royer, L. Ohler, T.A. Shaler, J.M. Hunter, D.J. Pollart, J.A. Monforte, and C.H. Becker. 1999. Single nucleotide polymorphism determination using primer extension and time-of-flight mass spectrometry. *Electrophoresis* 20(6):1258–1265.

Lindqvist, A.-K.B., P.K.E. Magnusson, J. Balciuniene, C. Wadelius, E. Lindholm, M.E. Alarcon-Riquelme, and U.B. Gyllensten. 1996. Chromosome-specific panels of tri- and tetranucleotide microsatellite markers for multiplex fluorescent detection and automated genotyping: evaluation of their utility in pathology and forensics. *Genome Research* 6:1170–1176.

MITOMAP: A Human Mitochondrial Genome Database. Center for Molecular Medicine, Emory University, Atlanta, GA, USA. *http://www.gen.emory. edu/mitomap.html.* 1999.

Monforte, J.A., C.H. Becker, T.A. Shaler, and D.J. Pollart. 1997. Oligonucleotide sizing using immobilized cleavable primers. U.S. Patent No. 5,700,642.

Monforte, J.A., T.A. Shaler, Y. Tan, and C.H. Becker. 1999. Methods of preparing nucleic acids for mass spectrometric analysis. U.S. Patent No. 5,965,363.

Ross, P.L. and Belgrader, P. 1997 Analysis of short tandem repeat polymorphisms in human DNA by matrix-assisted laser desorption/ionization mass spectrometry. *Analytical Chemistry* 69(19):3966–3972.

REFERENCES

Ross, P., L. Hall, I. Smirnov, and L. Haff. 1998. High level multiplex genotyping of MALDI-TOF mass spectrometry. *Nature Biotechnology* 16:1347–1351.

Rozen, S., and H.J. Skaletsky. 1998. Primer3. Code available at *http://www. genome.wi.mit.edu/genome_software/ other/primer3.html.*

Ruitberg, C.M., D.J. Reeder, and J.M. Butler. 2001. STRBase: A short tandem repeat DNA database for the human identity testing community. *Nucleic Acids Research* 29(1):320–322.

Schumm, J.W., A.W. Lins, K.A. Micka, C.J. Sprecher, D.R. Rabbach, and J.W. Bacher. 1997. Automated fluorescent detection of STR multiplexes— development of the GenePrint™ PowerPlex™ and FFFL multiplexes for forensic and paternity applications. In *Proceedings from the Seventh International Symposium on Human Identification 1996.* Madison, WI: Promega Corporation, 70–88.

Shaler, T.A., J.N. Wickham, K.A. Sannes, K.J. Wu, and C.H. Becker. 1996. Effect of impurities on the matrix-assisted laser desorption mass spectra of single-stranded oligonu- cleotides. *Analytical Chemistry* 68(3):576–579.

Shuber, A.P., V.J. Grondin, and K.W. Klinger. 1995. A simplified procedure for developing multiplex PCRs. *Genome Research* 5:488–493.

Sullivan, K.M., A. Mannucci, C.P. Kimpton, and P. Gill. 1993. A rapid and quantitative DNA sex test: fluorescence-based PCR analysis of X-Y homologous gene amelogenin. *BioTechniques* 15:637–641.

Taranenko, N.I., V.V. Golovlev, S.L. Allman, N.V. Taranenko, C.H. Chen, J. Hong, and L.Y. Chang. 1998. Matrix- assisted laser desorption/ionization for short tandem repeat loci. *Rapid Communications in Mass Spectrometry* 12(8):413–418.

Tully, G., K.M. Sullivan, P. Nixon, R.E. Stones, and P. Gill. 1996. Rapid detection of mitochondrial sequence polymorphisms using multiplex solid- phase fluorescent minisequencing. *Genomics* 34(1):107–113.

Underhill, P.A., L. Jin, A.A. Lin, S.Q. Mehdi, T. Jenkins, D. Vollrath, R.W. Davis, L.L. Cavalli-Sforza, and P.J. Oefner. 1997. Detection of numerous Y chromosome biallelic polymorphisms by denaturing high-performance liquid chromatography. *Genome Research* 7(10):996–1005.

Vestal, M.L., P. Juhasz, and S.A. Martin. 1995. Delayed extraction matrix-assisted laser desorption time- of-flight mass spectrometry. *Rapid Communications in Mass Spectrometry* 9:1044–1050.

Walsh, P.S., N.J. Fildes, and R. Reynolds. 1996. Sequence analysis and character- ization of stutter products at the tetranucleotide repeat locus vWA. *Nucleic Acids Research* 24(14): 2807–2812.

Wilson, M.R., D. Polanskey, J.M. Butler, J.A. DiZinno, J. Replogle, and B. Budowle. 1995. Extraction, PCR amplification, and sequencing of mito- chondrial DNA from human hair shafts. *BioTechniques* 18:662–669.

Wu, K.J., T.A. Shaler, and C.H. Becker. 1994. Time-of-flight mass spectrome- try of underivatized single-stranded DNA oligomers by matrix-assisted laser desorption. *Analytical Chemistry* 66:1637–1645.

Wu, K.J., A. Steding, and C.H. Becker. 1993. Matrix-assisted laser desorp- tion time-of-flight mass spectrome- try of oligonucleotides using 3-hydroxypicolinic acid as an ultraviolet-sensitive matrix. *Rapid Communications in Mass Spectrometry* 7:142–146.

PUBLISHED PAPERS AND PRESENTATIONS

From 1997 to 1999, six publications resulted from the work funded by NIJ, and at least one more manuscript is in preparation. All articles were published in journals or for conference proceedings that are accessible and frequented by forensic DNA scientists to ensure proper dissemination of the information.

Butler, J.M., J. Li, T.A. Shaler, J.A. Monforte, and C.H. Becker. 1998. Reliable genotyping of short tandem repeat loci without an allelic ladder using time-of-flight mass spectrometry. *International Journal of Legal Medicine* 112 (1): 45–49.

Butler, J.M., J. Li, J.A. Monforte, C.H. Becker, and S. Lee. 1998. Rapid and automated analysis of short tandem repeat loci using time-of-flight mass spectrometry. In *Proceedings of the Eighth International Symposium on Human Identification 1997.* Madison, WI: Promega Corporation, 94–101.

Butler, J.M., K.M. Stephens, J.A. Monforte, and C.H. Becker. 1999. High-throughput STR analysis by time-of-flight mass spectrometry. In *Proceedings of the Second European Symposium on Human Identification 1998.* Madison, WI: Promega Corporation, 121–130.

Butler, J.M., and C.H. Becker. 1999. High-throughput genotyping of forensic STR and SNP loci using time-of-flight mass spectrometry. In *Proceedings of the Ninth International Symposium on Human Identification 1998.* Madison, WI: Promega Corporation, 43–51.

Li, J., J.M. Butler, Y. Tan, H. Lin, S. Royer, L. Ohler, T.A. Shaler, J.M.

Hunter, D.J. Pollart, J.A. Monforte, and C.H. Becker. 1999. Single nucleotide polymorphism determination using primer extension and time-of-flight mass spectrometry. *Electrophoresis* 20:1258–1265.

Butler, J.M. 1999. STR analysis by time-of-flight mass spectrometry. *Profiles in DNA* 2(3): 3–6.

In addition, one patent was submitted based on work funded by NIJ. This patent describes the PCR primer sequences used to generate smaller amplicons for 33 different STR loci along with representative mass spectrometry results. The sequences for multiple cleavable primers are also described, although this proprietary chemistry is the subject of U.S. Patent 5,700,642, which was issued in December 1997. The process of multiplexing STR loci by interleaving the alleles on a compressed mass scale is also claimed by U.S. Patent 6,090,558 (Butler et al., 2000).

During the course of this NIJ grant, research findings were presented to the forensic DNA community at the following scientific meetings:

◆ Eighth International Symposium on Human Identification (September 20, 1997)

◆ San Diego Conference, Nucleic Acid Technology: The Cutting Edge of Discovery (November 7, 1997)

◆ NIJ Research Committee (February 8, 1998)

◆ American Academy of Forensic Sciences (February 13, 1998)

◆ Southwest Association of Forensic Scientists DNA training workshop (April 23, 1998)

◆ California Association of Criminalists DNA training workshop (May 6, 1998)

◆ National Conference on the Future of DNA (May 22, 1998)

◆ Florida DNA Training Session (May 22, 1998)

◆ American Society of Mass Spectrometry (June 4, 1998)

◆ Second European Symposium on Human Identification (June 12, 1998)

◆ IBC DNA Forensics Meeting (July 31, 1998)

◆ Ninth International Symposium on Human Identification (October 8, 1998)

◆ Fourth Annual CODIS User's Group Meeting (November 20, 1998)

◆ NIJ Research Committee (February 15, 1999)

In addition, the authors participated in NIJ's "Technology Saves Lives" Technology Fair on Capitol Hill in Washington, D.C., March 30–31, 1998, an event that provided excellent exposure for NIJ to Congress. Here, one of the DNA sample preparation robots was demonstrated in the lobby of the Rayburn Building. In September 1998, an 11-minute video was also prepared to illustrate some of the advantages of mass spectrometry for high-throughput DNA typing.

About the National Institute of Justice

NIJ is the research and development agency of the U.S. Department of Justice and is the only Federal agency solely dedicated to researching crime control and justice issues. NIJ provides objective, independent, nonpartisan, evidence-based knowledge and tools to meet the challenges of crime and justice, particularly at the State and local levels. NIJ's principal authorities are derived from the Omnibus Crime Control and Safe Streets Act of 1968, as amended (42 U.S.C. §§ 3721–3722).

NIJ's Mission

In partnership with others, NIJ's mission is to prevent and reduce crime, improve law enforcement and the administration of justice, and promote public safety. By applying the disciplines of the social and physical sciences, NIJ—

- **Researches** the nature and impact of crime and delinquency.

- **Develops** applied technologies, standards, and tools for criminal justice practitioners.

- **Evaluates** existing programs and responses to crime.

- **Tests** innovative concepts and program models in the field.

- **Assists** policymakers, program partners, and justice agencies.

- **Disseminates** knowledge to many audiences.

NIJ's Strategic Direction and Program Areas

NIJ is committed to five challenges as part of its strategic plan: 1) **rethinking justice** and the processes that create just communities; 2) **understanding the nexus** between social conditions and crime; 3) **breaking the cycle** of crime by testing research-based interventions; 4) **creating the tools** and technologies that meet the needs of practitioners; and 5) **expanding horizons** through interdisciplinary and international perspectives. In addressing these strategic challenges, the Institute is involved in the following program areas: crime control and prevention, drugs and crime, justice systems and offender behavior, violence and victimization, communications and information technologies, critical incident response, investigative and forensic sciences (including DNA), less-than-lethal technologies, officer protection, education and training technologies, testing and standards, technology assistance to law enforcement and corrections agencies, field testing of promising programs, and international crime control. NIJ communicates its findings through conferences and print and electronic media.

NIJ's Structure

The NIJ Director is appointed by the President and confirmed by the Senate. The NIJ Director establishes the Institute's objectives, guided by the priorities of the Office of Justice Programs, the U.S. Department of Justice, and the needs of the field. NIJ actively solicits the views of criminal justice and other professionals and researchers to inform its search for the knowledge and tools to guide policy and practice.

NIJ has three operating units. The Office of Research and Evaluation manages social science research and evaluation and crime mapping research. The Office of Science and Technology manages technology research and development, standards development, and technology assistance to State and local law enforcement and corrections agencies. The Office of Development and Communications manages field tests of model programs, international research, and knowledge dissemination programs. NIJ is a component of the Office of Justice Programs, which also includes the Bureau of Justice Assistance, the Bureau of Justice Statistics, the Office of Juvenile Justice and Delinquency Prevention, and the Office for Victims of Crime.

To find out more about the National Institute of Justice, please contact:

National Criminal Justice Reference Service
P.O. Box 6000
Rockville, MD 20849–6000
800–851–3420
e-mail: *askncjrs@ncjrs.org*

To obtain an electronic version of this document, access the NIJ Web site
(http://www.ojp.usdoj.gov/nij).

If you have questions, call or e-mail NCJRS.

www.ingramcontent.com/pod-product-compliance
Lightning Source LLC
Chambersburg PA
CBHW081212170526
45165CB00009B/2798